Business Intelligence and Analytics in Small and Medium Enterprises

Manufacturing Design and Technology Series
Series Editor: J. Paulo Davim, University of Aveiro, Portugal

This series will publish high-quality references and advanced textbooks in the broad area of manufacturing design and technology, with a special focus on sustainability in manufacturing. Books in the series should find a balance between academic research and industrial application. This series targets academics and practicing engineers working on topics in materials science, mechanical engineering, industrial engineering, systems engineering, and environmental engineering as related to manufacturing systems, as well as professions in manufacturing design.

Technological Challenges and Management
Matching Human and Business Needs
Edited by Carolina Machado and J. Paulo Davim

Advanced Machining Processes
Innovative Modeling Techniques
Edited by Angelos P. Markopoulos and J. Paulo Davim

Management and Technological Challenges in the Digital Age
Edited by Pedro Novo Melo and Carolina Machado

Machining of Light Alloys
Aluminum, Titanium, and Magnesium
Edited by Diego Carou and J. Paulo Davim

Additive Manufacturing
Applications and Innovations
Edited by Rupinder Singh and J. Paulo Davim

Emotional Intelligence and Neuro-Linguistic Programming
New Insights for Managers and Engineers
Edited by Carolina Machado and J. Paulo Davim

For more information about this series, please visit: https://www.crcpress.com/Manufacturing-Design-and-Technology/book-series/CRCMANDESTEC

Business Intelligence and Analytics in Small and Medium Enterprises

Edited by

Pedro Novo Melo and Carolina Machado

CRC Press
Taylor & Francis Group
Boca Raton London New York

CRC Press is an imprint of the
Taylor & Francis Group, an **informa** business

CRC Press
Taylor & Francis Group
6000 Broken Sound Parkway NW, Suite 300
Boca Raton, FL 33487-2742

Printed on acid-free paper

International Standard Book Number-13 978-0-367-17388-3 (Hardback)

Library of Congress Cataloging-in-Publication Data

Library of Congress Cataloging-in-Publication Data
Names: Melo, Pedro, 1981- editor. | Machado, Carolina, 1965- editor.
Title: Business intelligence and analytics in small and medium enterprises / edited by Pedro Novo Melo and Carolina Machado.
Description: Boca Raton, FL : CRC Press/Taylor & Francis Group, 2020. | Series: Manufacturing design and technology series | Includes bibliographical references and index. | Summary: "Technological developments in recent years have been tremendous. This evolution is visible in companies through technological equipment, computerized procedures, and management practices associated with technologies. One of the management practices that is visible is related to business intelligence and analytics (BI&A). Concepts such as data warehouse, key performance Indicator (KPIs), data mining, and dashboards are changing the business arena. This book aims to promote research related to these new trends that open up a new field of research in the small and medium enterprises (SMEs) area"-- Provided by publisher.
Identifiers: LCCN 2019034898 (print) | LCCN 2019034899 (ebook) | ISBN 9780367173883 (hardback ; acid-free paper) | ISBN 9780429056482 (ebook)
Subjects: LCSH: Small business--Data processing. | Business intelligence.
Classification: LCC HD62.7 .B854 2020 (print) | LCC HD62.7 (ebook) | DDC 658.4/72--dc23
LC record available at https://lccn.loc.gov/2019034898
LC ebook record available at https://lccn.loc.gov/2019034899

Visit the Taylor & Francis Web site at
http://www.taylorandfrancis.com

and the CRC Press Web site at
http://www.crcpress.com

Contents

Preface

Technological developments in recent years have been tremendous. Technology is increasingly present in several dimensions of society. This evolution is visible in organizations through increasingly technological equipment, increasingly computerized procedures and management practices associated with technologies. One of the management practices that are visible is related to business intelligence and analytics (BI&A).

Concepts such as data warehouse, KPIs (Key Performance Indicators), Data mining, and dashboard are changing the Business area. BI&A are a set of methodologies, processes, architectures, and technologies that allow to give meaning to the existing data in the different kind of organizations, making them relevant for the decision-making. Taking into account these challenges, this book aims to promote research and the knowledge transmission related to these new trends that open up a new field of research in the SMEs area.

This book focuses on the latest and more recent research findings that are occurring in this field in different countries. More specifically, it looks to

- show in what ways organizations around the world are facing today's technological challenges;
- share knowledge and insights on an international scale;
- help researchers and practitioners to select among the different options and strategies the more relevant priorities to manage competitive organizations; and
- keep the readers and researchers informed about the latest developments in the field and of forthcoming international studies.

For the purpose of sharing knowledge, through debate and information exchange, about technological challenges and management in digital age, this book is divided into nine chapters: Chapter 1 covers "Process Mining – Prerequisites and Their Applicability for SMEs". Chapter 2 discusses "Using Customer Analytics to Succeed: The Case of Mexican SMEs". Chapter 3 contains "Data Management Software Solutions for Business Sustainability – An Overview". Chapter 4 describes "A Paradigm Shift in Accounting and Auditing in the *Era* of Big Data". Chapter 5 covers "Mobile Advertising Framework: Format, Location, and Context". Chapter 6 describes "Marketing Analytics. Why Measuring Web and Social Media Matters". Chapter 7 discusses "Managers' Perception of Business Intelligence Capability of SMEs in Turkey". Chapter 8 contains "The Development of Loyalty Programs in the Retail Sector". Finally, in Chapter 9 "Business Intelligence, Big Data and Data Analysis" is discussed.

Serving as a useful reference for academics, researchers, managers, engineers, and other professionals in related matters with technologies, management, and engineering innovation, this book, entitled *Business Intelligence and Analytics* looks to addresses several dimensions of technology in favor of management with a particular emphasis on business intelligence and analytics, and its impact on business and the organization competitiveness.

The Editors are grateful to CRC Press/Taylor & Francis Group for this opportunity and for their professional support. Finally, we thank all chapter contributors for their interest and time allotted to work on this project.

Pedro Novo Melo
Barcelos, Portugal

Carolina Machado
Braga, Portugal

Editors

Pedro Novo Melo is assistant professor at the Polytechnic Institute of Cávado and Ave. He has a PhD degree in Business Administration from the University of Minho. His main areas of interest are organizational behavior, human resource practices and its relationship with the business strategy, with special emphasis on SMEs. In recent years, he has focused his research on digital HRM and Innovation in human resources management.

Carolina Machado received her PhD degree in Management Sciences (Organizational and Politics Management area/ Human Resources Management) from the University of Minho in 1999, Masters degree in Management (Strategic Human Resource Management) from the Technical University of Lisbon in 1994, and Degree in Business Administration from the University of Minho in 1989. Teaching in the Human Resources Management subjects since 1989 at University of Minho, she is an associated professor since 2004, with experience and research interest areas in the field of Human Resource Management, International Human Resource Management, Human Resource Management in SMEs, Training and Development, Emotional Intelligence, Management Change, Knowledge Management, and Management/HRM in the Digital Age. She is head of the Department of Management and head of the Human Resources Management Work Group at University of Minho, as well as chief editor of the *International Journal of Applied Management Sciences and Engineering* (IJAMSE), guest editor of journals, books editor and book series editor, as well as reviewer in different international prestigious journals. In addition, she has also published, both as editor/coeditor and as author/coauthor, several books, book chapters, and articles in journals and conferences.

Contributors

Bilal Aslam
School of Business and Economics
University of Jyväskylä
Jyväskylä, Finland

Christopher Bernhard
Salzburg University of Applied Sciences
Puch/Salzburg, Austria

Davide Carneiro
School of Management and Technology
Polytechnic Institute of Porto
Porto, Portugal

Luís Ferreira
Hight School of Technology
Researcher – 2Ai – Applied Artificial Intelligence Laboratory
Polytechnic Institute of Cávado and Ave
Barcelos, Portugal

Mine Afacan Findikli
Department of Business Management
Beykent University
İstanbul, Turkey

Alexandra Fontes
Research on Economics, Management and Information Technologies (REMIT)
Polytechnic Institute of Viana do Castelo
Viana do Castelo, Portugal

Heikki Karjaluoto
School of Business and Economics
University of Jyväskylä
Jyväskylä, Finland

Xabier Martínez-Rolán
Faculty of Social Science and Communication
University of Vigo
Pontevedra, Spain

Adelaide Martins
Research on Economics, Management and Information Technologies (REMIT)
University Portucalense
Porto, Portugal

Mihaela-Filofteia Tutunea
Faculty of Business
Babeş-Bolyai University
Cluj-Napoca, Romania

Flor Morton
Management Department
Universidad de Monterrey
Monterrey, Mexico

Julian Marius Müller
Salzburg University of Applied Sciences
Puch/Salzburg, Austria

Teresa Piñeiro-Otero
University of Coruña
Coruña, Spain

Hélder Quintela
Hight School of Technology
Polytechnic Institute of Cávado and Ave
Barcelos, Portugal

Muniba Rauf
School of Business and Economics
University of Jyväskylä
Jyväskylä, Finland

Ana Paula Silva
Polytechnic Institute of Viana do Castelo
Viana do Castelo, Portugal

Mustafa Sundu
Department of Business Management
Beykent University
İstanbul, Turkey

Teresa Treviño
Management Department
Universidad de Monterrey
Monterrey, Mexico

Okan Yaşar
Beykent University
İstanbul, Turkey

Laura Zapata-Cantú
Strategy and Management Department
EGADE Business School
Tecnológico de Monterrey
Monterrey, Mexico

Alexander Zeisler
Salzburg University of Applied Sciences
Puch/Salzburg, Austria

1 Process Mining – Prerequisites and Their Applicability for Small and Medium-sized Enterprises

Alexander Zeisler
Salzburg University of Applied Sciences, Puch/Salzburg, Austria

Christopher Bernhard
Salzburg University of Applied Sciences, Puch/Salzburg, Austria

Julian Marius Müller
Salzburg University of Applied Sciences, Puch/Salzburg, Austria

CONTENTS

1.1 INTRODUCTION

Rapid technical progress, the increasing capability to generate and store data, and the growing fusion of the physical and the digital world are fostering the use of new technologies (Accorsi et al., 2012). Process Mining is a trend in business process management which has finally arrived in industrial application. The most powerful driver for process mining is the collaboration between process mining providers and vendors of enterprise applications such as SAP, Oracle, or Salesforce. Large companies like Siemens or Vodafone could already realize significant benefits by applying process mining tools in business process management (Kerremans, 2018). The reduction of throughput times, cutting costs, or increased satisfaction among customers are examples how an organization can benefit from process mining in order to remain competitive or to gain further competitive advantages.

It is evident that small- and medium-sized enterprises (SMEs)[1] show a considerable backlog when approaching the digital transformation compared to large enterprises. Only a small proportion of SMEs is already prepared to foster the full potentials, whereas the majority of SMEs is beginning with initial test applications (Müller, 2019; Müller et al., 2018). The digitization of the value chain, a key aspect in the context of a digitized industrial value chain, is also described as the concept of Industry 4.0 (Müller et al., 2018). It is further the fundamental basis for process mining. In the course of an INTERREG-funded project,[2] companies from the federal state of Bavaria (Germany) and the federal state of Salzburg (Austria) were being asked to participate in a digital readiness check. This readiness check is focusing on digitization of the value chain, and the result reaffirms the outcome of the above-mentioned study "Digitization in SMEs 2018". The level of digitization of the value chain is low; SMEs self-evaluate the digitization of their value chain with a value of 2.1 on average on a scale ranging from 0 to 6 points. Another distinguishing feature between SMEs and large enterprises is the assumption that large enterprises usually have implemented professional business process management and are performing work on the basis of explicit and formal processes (Müller et al., 2018; Müller and Voigt, 2018). SMEs, on the other hand, are expected to largely perform the activities on an implicit basis and professional business process management is not necessarily established (Burattin, 2015).

Based on these characteristics of SMEs described in extant literature, two core questions are of interest: The first question is about the prerequisites for process mining and when an organization is ready for this technology. The second question deals with the applicability of process mining for SMEs. It has to be evaluated if process mining is only suitable for large companies, or if similar benefits can be realized in SMEs – under given circumstances that a low level of digitization has to be expected.

To find an answer to these questions, two steps are conducted: After explaining the basic functionalities of process mining, prerequisites for successful process mining are presented by combining both qualitative and quantitative research methods. These findings are then tested in case studies with two SMEs, both suppliers for the automotive industry and both situated in the Austrian-German border region. The case studies have been carried out in cooperation between the respective SME and the researchers at Salzburg University of Applied Sciences.

1.2 WHAT IS PROCESS MINING?

In short, process mining "[...] uses business process events for process visualization and analytics" (Yli-Pietilä and Kauppinen, 2016).

A process can be defined as a "sequence of activities performed in a specific order to achieve a specific goal" (Munoz-Gama, 2016). End-to-end processes – like order-to-cash, manufacturing processes, or service processes – are integral part of industries and professional business process management is crucial for companies who want to be competitive in an ever faster and complex economic environment. Process mining is a relatively young research discipline and is bridging the gap between process science and data science, aiming to discover, monitor, and improve real processes by using data from event logs (Van der Aalst, 2016). Today, business processes are being performed with support of IT systems to varying degrees and thus, process mining is possible due to the simple fact that data already exists (Rozinat and Günther, 2014). Information about business processes is being extracted from enterprise transaction systems and hence, information about real-life processes can be generated on the basis of data-driven facts (Davenport and Spanyi, 2019). Process mining offers an innovative approach to analyze the performance of a process. Commonly used manual tools – like spreadsheets in Excel, dashboards, or Power Point slides – are being replaced by dynamic tools. Process mining tools are visually reconstructing the actual flow of business processes, which helps to create a common understanding and process transparency among an organization. As a result, process analysis can be performed much more quickly and efficiently compared to the manual approach (Rozinat and Günther, 2014).

1.3 PREREQUISITES FOR SUCCESSFUL PROCESS MINING

Process mining is a tool to professionalize business process management and needs to be embedded in an adequate professional environment. An organization needs to be prepared accordingly if it wants to implement a process mining tool. The fundamental requirements and prerequisites for successful process mining have been elaborated in the scope of an empirical analysis. First, seven guided interviews have been conducted with experts from three relevant areas: experts from the academic environment, business process experts from the industry, and providers of process mining tools. The outcome of these expert interviews laid the foundation for a quantitative survey that has been carried out in a second step among companies in the German-speaking area. Seventy-nine valid responses have been received from the quantitative survey and as seen in Figure 1.1, both – large enterprises (66%) but also SMEs (34%) – provided responses.

The results have been consolidated in seven prerequisites that are recommended being fulfilled before implementing a process mining tool. These prerequisites are of a general character and can be applied for all types of enterprises.

1.3.1 Organizational Prerequisites

Implementing a new technology – like process mining – requires full support from the management team. That does not necessarily implicate that the management

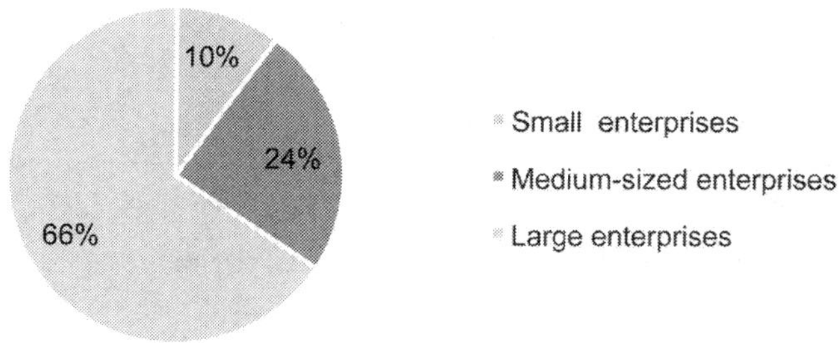

FIGURE 1.1 Categories of participating companies.

team is responsible for the implementation itself, but an adequate environment in the enterprise needs to be created. The management team is accountable to define setup in the organization (e.g., create an own department in Data and Quality Management). A project manager needs to be appointed and all involved employees need to be empowered according to their tasks. Furthermore, the management team has to assure that the implementation of the new technology is understood, accepted, and being supported from all involved functions in the organization. Companies should further truthfully inform employees about the benefits and the expected outcome of process mining. Otherwise, there is a risk that employees might block and put up resistance against process mining, especially if they feel being systematically controlled. The organizational prerequisites are crucial to implement process mining – as they are for any new technology, or for organizational changes in general.

1.3.2 Process-Related Prerequisites

Companies are advised to change their culture from thinking in departmental structures to creating a holistic understanding of end-to-end processes. Being aware of the relevance and impact of business processes is another prerequisite for successful process mining. A deep understanding of cross-divisional and companywide processes among employees is the fundamental basis for this. It is recommended that processes are described and depicted in a consistent way, ideally by using business process management software. The awareness about efficient processes management will have a positive impact on customer satisfaction, compliance, and the quality of its products. It is essential to note, that a formal certification – like ISO 9001 – per se does not necessarily fulfill this requirement in a sufficient manner. Such certifications are often carried out due to legal requirements or requirements from customers, but the certification does not necessarily assure that holistic end-to-end business process management is established in the organization.

1.3.3 IT-Related Prerequisites

It is essential that processes are being executed with support of IT systems (Enterprise Resource Planning – Enterprise Resource Planning (ERP) Systems, planning tools, or logistics software) to a large extent. Activities need to "leave digital tracks". That does not necessarily mean that 100% of the processes need to be fully digitized, as long as the person that conducts process mining (process miner) is aware of black boxes[3] in the process. In case that the majority of activities does not leave any or only minimal digital tracks, the company is recommended to strive for digitization first, as process mining can support to identify those black boxes in case of poorly digitized processes. The process miner further needs to be aware that process mining might provide very little benefits in case of poorly digitized processes and that there is a certain risk that even a wrong understanding of the process can be created. There is no generally applicable recommendation about an appropriate degree of digitization of a process, each process needs to be investigated individually. In the best case, there is a companywide ERP system established, and the processes are fully digitized within a single leading system.

1.3.4 Data-Related Prerequisites

In combination with IT-related prerequisites, it is indispensable that process-related data is available in adequate quality. An analysis can only be as good as the quality of its data. Process-related data need to fulfill three minimum requirements: Each event has to refer to a case (i.e., single process instance), an activity (i.e., a well-defined step in the process), and a time stamp (Van der Aalst, 2011). A unique case ID is required to assign all events to the respective case. Any type of master data can be included into the event log in addition, to enable profound analysis of the process (Van der Aalst, 2016). Table 1.1 shows a fragment of a representative event log. All events in this example refer to a Case ID, an activity, and they have a time stamp. Information about resource and cost are included as supplementary master data.

Given the fact that these three requirements are fulfilled, the quality of data needs to be examined in addition. Van der Aalst et al. (2016) propose five maturity levels for event logs in order to systematically assess the quality of data and its applicability for process mining. The maturity levels range from the lowest level, where data is of poor quality (e.g., manual documentation paper documents) up to the highest level, where data is of excellent quality. Events are well defined and recorded

TABLE 1.1
A Fragment of Some Event Log

Case ID	Event ID	Time Stamp	Activity	Resource	Cost
1	35654423	30-12-2010:11.02	Register request	Pete	50
	35654424	31-12-2010:10.06	Examine thoroughly	Sue	400
	35654425	05-01-2011:15.12	Check ticket	Mike	100
	35654426	06-01-2011:11.18	Decide	Sara	200
	35654427	07-01-2011:14.24	Reject request	Pete	200
2	35654483	32-12-2010:11.32	Register request	Mike	50
	35654485	30-12-2010:12.12	Check ticket	Mike	100
	35654487	30-12-2010:14.16	Examine casually	Pete	400

Source: Based on Van der Aalst (2016).

automatically, systematically, reliable, and safe in the highest maturity level (Van der Aalst et al., 2016). Maturity levels are described in Table 1.2.

1.3.5 Employee-Related Prerequisites

The process miner needs to fulfill two criteria: He should have a profound understanding of processes and systems and he needs to have a high affinity for new technologies.

TABLE 1.2
Maturity Level for Event Logs

Level	Characterization	Examples
*****	Highest level: The event log is of excellent quality (i.e., trustworthy and complete) and events are well-defined. Events are recorded in an automatic, systematic, reliable, and safe manner. Privacy and security considerations are addressed adequately. Moreover, the events recorded (and all of their attributes) have clear semantics. This implies the existence of one or more ontologies. Events and their attributes point to this ontology.	Semantically annotated logs of BPM systems.
****	Events are recorded automatically and in a systematic and reliable manner, that is, logs are trustworthy and complete. Unlike the systems operating at level, notions such as process instance (case) and activity are supported in an explicit manner.	Events logs of traditional BPM/workflow systems.
***	Events are recorded automatically, but no systematic approach is followed to record events. However, unlike logs at level, there is some level of guarantee that the events recorded match reality (i.e., the event log is trustworthy but not necessarily complete). Consider, for example, the events recorded by an ERP system. Although events need to be extracted from a variety of tables, the information can be assumed to be correct (e.g., it is safe to assume that a payment recorded by the ERP actually exists and vice versa).	Tables in ERP systems, event logs of CRM systems, transaction logs of messaging systems, event logs of high-tech systems, etc.
**	Events are recorded automatically, that is, as a by-product of some information system. Coverage varies, that is, no systematic approach is followed to decide which events are recorded. Moreover, it is possible to bypass the information system. Hence, events may be missing or not recorded properly.	Event logs of document and product management systems, error logs of embedded systems, worksheets of service engineers, etc.
*	Lowest level: Event logs are of poor quality. Recorded events may not correspond to reality and events may be missing. Event logs for which events are recorded by hand typically have such characteristics	Trails left in paper documents routed through the organization (yellow notes), paper-based medical records, etc.

Source: Adapted from Van der Aalst et al. (2016).

As already mentioned, an organization should establish holistic end-to-end process management. Process miners thus need to be capable to understand such complex processes in a way that they are able to develop a feeling for requirements and tasks of all parties involved. Furthermore, they need to recognize connections and dependencies between all affected functions. The task of a process miner is to identify, prepare, and provide the relevant systems and all data that is required for the analysis.

The necessary depth of specific knowledge about process mining is largely depending on the strategy an enterprise is pursuing, as there is a wide range of solutions being offered from software providers. Process mining can be implemented autonomously, with support of consulting partners, or even the whole analysis can be outsourced to a partner company (e.g., software providers).

1.3.6 Legal Requirements

Special attention needs to be paid to data protection and security, as personnel data is being logged in various IT systems. Sensitive data – with special focus to personal data – needs to be protected and it has to be assured that all legal requirements are fully respected. Real names from employees, suppliers, or customers can be mentioned as an example for sensitive data. It has to be assured that it is impossible to identify a real person within a process mining analysis. This requirement can be fulfilled by applying appropriate algorithms (anonymization and pseudonymization). It is strongly recommended to involve the works council at an early stage in order to facilitate a smooth implementation of the new technology.

1.3.7 Means and Resources

Implementing process mining requires individual planning of means and resources. Financial resources, personnel resources, and time need to be assessed carefully. As mentioned earlier, a broad range of software and service providers is offering various process mining solutions on the market. Financial planning should consider costs for the process mining tool, for the licenses, for consulting (if required), for training, and for personnel resources. Yearly costs can range from some thousand euros up to several million euros per year – depending on the size of the company an on the implementation strategy. The main cost driver for implementing process mining can be costs for consulting.

1.4 PROCESS MINING IN SME – TWO CASE STUDIES

The proposed prerequisites for process mining are of a general character and it needs to be figured out if and how they are applicable for SMEs. Two individual case studies in SMEs in the German-Austrian border region have been carried out, aiming to answer two guiding questions:

- Is process mining a suitable technology for SMEs?
- Are the seven prerequisites for process mining being fulfilled in the respective SME?

The L*lifecycle model, a model recommended by Van der Aalst et al. (2011) for process mining projects of Lasagna processes,[4] was chosen as a guiding principle. The L*lifecycle model describes five stages that shall be performed within a process mining project. The most important stages in this model are – as they are in any other project related to big data – stages 0, named "plan and justify" and stage 1, named "extract". Exploratory data analysis happens in these early stages and it is crucial to understand both – business and data (Yu, 2017). At the beginning of a process mining project, the process needs to be selected and the problems need to be addressed. It is of major importance to understand the IT landscape and all relevant functions involved. A deep knowledge about available data and their quality has to be created. Regarding the two case studies, most effort has been spent in stage 0, including an analysis if the identified seven prerequisites are fulfilled. The following processes have been chosen for further analysis:

- Case Study 1: Production planning
- Case Study 2: Delivery process

Since the scope has been defined, a deep understanding of business and data has been elaborated. The processes have therefore been analyzed in three dimensions: process flow, landscape of systems/tools (analogue and digital) and involved functions. Both companies are using an ERP system, which was the main source for data.

Each of the seven prerequisites has been examined thoroughly and has been graded on a scale from 0 points (prerequisite not fulfilled at all) up to 10 points (prerequisite fully met). The examination has been proposed by the researchers from Salzburg University of Applied Sciences and subsequently presented to and aligned with representatives from the respective enterprise. The outcome of this analysis was a mostly equivalent result for both case studies. The outcome of case study 1 is illustrated in Figure 1.2.

Although the majority of prerequisites have been broadly fulfilled, it turned out that process mining would add only limited value in both processes at this point

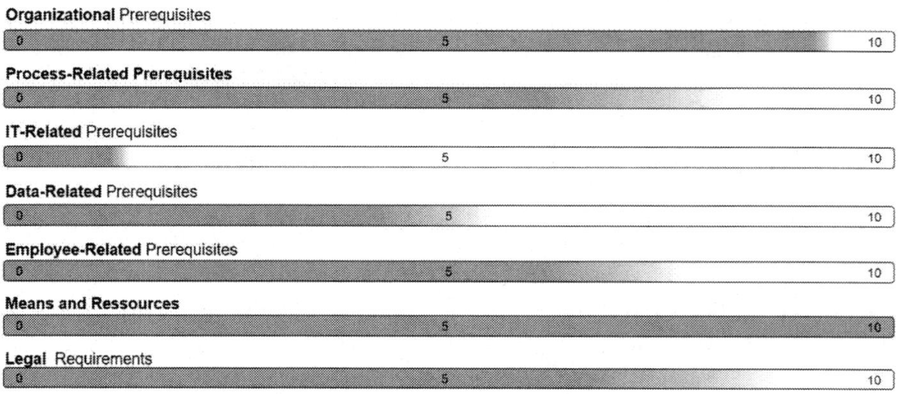

FIGURE 1.2 Prerequisites Case Study 1. (Own representation.)

in time. Even if organizational prerequisites, employee-related prerequisites, means and resources, and legal requirements are in good shape, meaningful use of process mining mainly depends on the following three requirements:

- Process-related prerequisites
- IT-related prerequisites
- Data-related prerequisites

These three requirements have been identified as the key-prerequisites for successful implementation of process mining in the course of the case studies. Although there was a good understanding of the process given, it turned out that the respective process was only marginally supported by IT systems, although both SMEs are using an ERP system. The reason behind is that the full functionality of the ERP system is not being exploited for the time being. The majority of activities is being done manually and communication happens on the short way (via telephone or personally). In Case Study 1, 6 of 18 activities in the standard process fulfilled the minimum requirements for a meaningful event log (case ID, activity, timestamp). At the end, it was decided together with each SME to stop the process mining project at this stage.

Nevertheless, valuable findings have been gained in these two case studies. It has become evident that both companies are not yet ready for valuable use of process mining. They have been recommended to push forward digitization of their key processes. By applying the L*lifecycle model and a deep analysis of the as-is process, black boxes have been identified and this analysis is providing valuable input for continuous improvement. Both companies confirmed that process mining will bring valuable input for their business in a future point in time.

Summarizing, the two leading questions have been answered as follows.

1.4.1 Is Process Mining a Suitable Technology for SMEs?

Process mining can bring added value to an SME, but the SME needs to be prepared diligently. Hereby, the typical restrictions of SMEs regarding digitization need to be respected, especially IT-related competencies, financial resources, and lacking economies of scale, among further (Müller 2019; Müller et al., 2018).

1.4.2 Are the Seven Identified Prerequisites for Process Mining Being Fulfilled in the Respective SME?

In both cases, the majority of prerequisites has been fulfilled, but there was a significant lack in the leading prerequisites: process-related prerequisites, IT-related prerequisites, and data-related prerequisites.

1.5 CONCLUDING REMARKS

Process mining is a new technology that creates high expectations in industrial application. Companies are expecting to significantly improve efficiency in operations and reporting, transparency, and common understanding through increased professional business management. Seven prerequisites are recommended to being fulfilled before starting process mining. Through conducting two case studies with SMEs, three key prerequisites have been identified as follows:

- Process-related prerequisites
- IT-related prerequisites
- Data-related prerequisites

From a managerial perspective, SMEs as well as large enterprises can benefit from process mining equally, provided that the key prerequisites are fulfilled in a sufficient way. In this regard, this article presents first insights regarding prerequisites that need to be fulfilled in order to implement process mining. Further, two case studies serve as an empirical basis to better understand the implementation process of process mining within SMEs. Thereby, this chapter adds to sparse empirical evidence on the unfolding of Industry 4.0 in SMEs in a less regarded technology (Müller 2019; Müller et al., 2018). For future research, the implementation of process mining should be regarded not only from the perspective of SMEs or large enterprises, but regarding their interplay when implementing process mining in an entire supply chain.

NOTES

1. Definition of SMEs: 1–250 employees, annual turnover up to 50 mEUR. Based on the definition of the European Commission (Liikanen, 2003).
2. Interreg Österreich – Bayern 2014 – 2020. KMU 4.0 – Digitaler Mittelstand. www.kmu40.eu
3. For example, manual activities and documents that do not leave digital tracks.
4. A Lasagna process has a clear structure and most cases are handled in a prearranged manner. *Source*: Van der Aalst (2016).

REFERENCES

Accorsi, R., Ullrich, Meike, Van der Aalst, Will. 2012. Process mining. Informatik-Spektrum, 35(5), 354–359.

Burattin, A. 2015. Process Mining Techniques in Business Environments. Theoretical Aspects, Algorithms, Techniques and Open Challenges in Process Mining. Lecture Notes in Business Information Processing. http://dx.doi.org/10.1007/978-3-319-17482-2.

Davenport, T. H., Spanyi, A. 2019. What Process Mining Is and Why Companies Should Do it. Harvard Business Review. https://hbr.org/2019/04/what-process-mining-is-and-why-companies-should-do-it (accessed May 8, 2019).

Kerremans, M. 2018. Market Guide for Process Mining. https://www.gartner.com/doc/reprints?id=1-4VX2Z7N&ct=180412&st=sb (accessed May 21, 2019).

Liikanen, E. 2003. Commission Recommendation of 6 May 2003 concerning the definition of micro, small and medium-sized enterprises. European Commission. https://eur-lex.europa.eu/legal-content/EN/TXT/PDF/?uri=CELEX:32003H0361&from=EN (accessed May 16, 2019).

Müller, J. M. 2019 Business model innovation in small- and medium-sized enterprises: Strategies for industry 4.0 providers and users. Journal of Manufacturing Technology Management, available online.

Müller, J. M., Buliga, O., Voigt, K.-I. 2018. Fortune favors the prepared: How SMEs approach business model innovations in Industry 4.0. Technological Forecasting and Social Change, 132, 2–17.

Müller, J. M., Voigt, K.-I. 2018. Sustainable industrial value creation in SMEs: A comparison between industry 4.0 and made in China 2025. International Journal of Precision Engineering and Manufacturing-Green Technology, 5(5), 659–670.

Munoz-Gama, J. 2016. Conformance Checking and Diagnosis in Process Mining. Comparing Observed and Modeled Processes. Lecture Notes in Business Information Processing. http://dx.doi.org/10.1007/978-3-319-49451-7.

Rozinat, A., Günther, C. 2014. The added value of process mining. BP Trends. https://www.bptrends.com/the-added-value-of-process-mining/ (accessed May 21, 2019).

Van der Aalst, W. 2011. Process Mining: Discovery, Conformance and Enhancement of Business Processes. Heidelberg: Springer.

Van der Aalst, W. 2016. Process Mining. Data Science in Action (2nd ed.). http://dx.doi.org/10.1007/978-3-662-49851-4.

Yli-Pietilä, M., Kauppinen, J. 2016. Process mining belongs to industrial digitalization. Midagon Industrial Digitalization White Paper. Midagon. https://www.midagon.com/app/uploads/2016/07/Midagon-White-Paper-PDE-20160926.pdf (accessed May 21, 2019).

Yu, C. H. 2017. Exploratory data analysis. Oxford Bibliographies. https://www.oxfordbibliographies.com/view/document/obo-9780199828340/obo-9780199828340-0200.xml?rskey=u8inUi&result=1&q=exploratory+data#firstMatch (actualized November 29, 2017, accessed May 20, 2019).

2 Using Customer Analytics to Succeed: The Case of Mexican SMEs

Laura Zapata-Cantú
Strategy and Management Department, EGADE Business School, Tecnológico de Monterrey, Monterrey, Mexico

Teresa Treviño
Management Department, Universidad de Monterrey, Monterrey, Mexico

Flor Morton
Management Department, Universidad de Monterrey, Monterrey, Mexico

CONTENTS

2.1 INTRODUCTION

To create a competitive advantage, companies have to think of new ways to improve their businesses. The adoption of business intelligence (BI), platforms, tools, and solutions seems to be an excellent business strategy that has become an essential component of contemporary business decision support systems. One benefit of BI is the opportunity to generate value for the firm when translation of big data to competitive advantages is well performed. This has important implications for managers who have to identify those digital strategies that could be competitive now and in the future. This strategy seems to be exclusive for large companies that require more storage technology based on the need to increase information storage capacities, but also could be a great opportunity for small- and medium-sized enterprises (SMEs) to succeed.

Even when among SMEs flexibility and a faster response are advantages, there is a general lack of technical expertise required for the conversion of data into information to facilitate an informed business decision-making process. Furthermore, a lack of understanding of the benefits of BI and budget also makes SMEs reluctant to invest in adopting any BI solutions (Raj, Wong, & Beaumont, 2016). This situation has prevailed because, generally, large companies have more assets in information technologies; SMEs need to know what suppliers, customers, and stakeholders' want, not only to be informed (or in communication) with one another but also to make strategic decisions. Similar to large companies, SMEs need to remember how well they need to be connected to respond in a faster manner to all their stakeholders.

The competitiveness of SMEs depends on their ability to establish mechanisms that, in a continuous way, allow them to acquire shares and apply data, information, and knowledge. BI platforms and tools allow companies to cope with the complexities and changes in the current uncertain environment. Focusing on Latin America, there is a low investment rate in technological platforms compared with higher growth emerging economies (McKinsey Report, 2019). The region has not been at the frontier of past and current innovation changes, mainly due to limited incentives to innovate. In the case of the SMEs in Mexico, there are many organizations wherein the information available for decision-making is often misleading, inaccurate, and tardy (Fernandez et al., 2016). Mexican SMEs' growth and management are based on the owners' expertise that exposes several problems summarized as a lack of strategic planning and lack of professionalization of the core business areas. In this regard, BI is not a priority even when it could be a solution to having operational efficiency, process control, and a customer orientation.

For instance, the use of social media analytics with a large online community enables SMEs to better reach their customers at an unprecedented scale and lower costs (Kurniawati, Shanks, & Bekmamedova, 2013). There is a widespread adoption of popular social media platforms by users around the world, and social media is already a critical part of the information ecosystem. In this regard, the use of social media can represent for SMEs a first step for digital strategy providing an opportunity to gain information about the market, establish a two-way communication with customers, and finally, improve customer engagement (McKinsey, 2016a).

In this chapter, we provide an overview of the challenges and opportunities that Mexican SMEs are facing regarding BI and suggest how social media as customer analytics are starting to be a relevant strategy to succeed. Social media represents a fundamental change in how information is being produced, transferred, and consumed. People around the world are creating content—also known as user-generated content—which may serve as a connection between organizations and consumers. Companies and especially SMEs can benefit from tracking information on social media to gain feedback and insights in how to better serve consumers and succeed in this competitive market (Leskovec, 2011). In particular, social media analytics are a rapidly emerging capability that provides companies with the opportunity to analyze online content to find insights about the attitudes and behavior of the market.

This chapter is structured as follows. The next section presents the current relevance of BI and analytics (BI&A) for SMEs. Section 2.3 exposes the economical context of the SMEs in Mexico. Section 2.4 focuses on social media analytics tools specifically for customer engagement in SMEs. Finally, Section 2.5 exposes some challenges and opportunities for Mexican SMEs in the digital era.

2.2 BUSINESS INTELLIGENCE AND ANALYTICS IN SMEs

The term business analytics emerged in the late 2000s because of the need to focus on the analytical component of BI. BI refers to a broad range of methodologies, processes, architectures, technologies, and solutions that gather, consolidate, analyze, and provide access to real-time data and transform them into meaningful information that assists managers in making better business decisions. Presently, BI&A is used as a unified concept.

Big data is a crucial term for BI&A. Big data describes the huge volumes of data that need sophisticated methods and technologies for data management and analytics (Iqbal et al., 2018). BI&A applies statistical, processing, and analytic techniques to big data for business decisions (Grover et al., 2018). Chiang et al. (2018) identify five characteristics of big data, namely, the volume (the amount of data), velocity (the frequency of data occurrence), variety (data types), veracity (clean data), and value (also called "data monetization"). According to these authors, generating value through data is the true contribution of big data to an organization, and the success of BI&A projects requires an understanding of how to translate big data into competitive advantages and strategic value for the firm. However, this is one of the key challenges for all businesses and researchers that use big data.

For instance, the HACE theorem explains the features of big data, which start with large-volume heterogeneous, autonomous sources with distributed and decentralized control, and seeks to explore complex and evolving relationships among data (Wu et al., 2013, p. 3). Big data is obtained through different information collectors that use their own schemata for data recording that result in heterogeneous representations of the same data; this heterogeneity represents a major challenge when attempting to aggregate data by combining the data from all sources. Additionally, each data source can generate and collect information without involving any centralized control, which protects big data applications such as Google and Facebook from attacks or malfunctions. These applications have a large number of server farms all over the world to ensure nonstop services and quick responses for local markets. However, protection from attacks and technical malfunctioning is not the only reason for having autonomous and decentralized sources; differences in legislation and regulation, seasonal promotions, top selling products, and customer behaviors in different countries/regions are also some of the reasons. Finally, the complexity and the relationships underneath the data increase as the volume of big data increases. People form friend circles based on the biological and social connections that commonly exist, not only in the physical space but also in the virtual world. These connections between individuals complicate the representation and reasoning process of big data applications that consider the complex data relationships and evolving changes to discover useful patterns from big data collections. Therefore, using big data is a current challenge for companies.

Big data is not only for large companies with large budgets; in fact, big data value represents an extremely strategic and profitable opportunity for SMEs, which can also benefit from massive amounts of online and offline information to make data-driven decisions, extend their products and services, and create new products and services to grow their businesses. Small-business owners should embrace the transition from an intuition-driven company to an analytics-driven company (Ogbuokiri, Udanor, & Agu, 2015). However, despite the benefits that big data represents for SMEs, many environmental factors are responsible for low level of big data adoption in these types of companies such as low level of understanding of big data, lack of infrastructure to analyze data, lack of in-house big data experts that results in reduced control over data and issues concerning data security, and lack of financial resources to access data analytics consulting, among other factors (Iqbal et al., 2018).

Considering the factors that limit SMEs' adoption of big data, the literature recommends several characteristics that SMEs should seek when implementing big data solutions. First, SMEs should seek flexibility. SMEs regularly employ a variety of solutions throughout the organization, which leads to an environment with many different types of data; therefore, the big data solution must allow these companies to choose only the capabilities that they need and to leverage the solutions and systems already in place without the need to replace them, as they are already implemented and adopted capabilities. Second, SMEs should seek simplicity. The solution should be easy to implement and use without the need for the IT department to become involved and should not need much training for a company to start using it in only a few days or weeks. Additionally, all the capabilities of the system should work together seamlessly and should allow the integration of new capabilities with existing systems without the need for expensive specialists. Third, SMEs should seek the appropriate cost. The solution must be priced right, and SMEs should be able to pay for only the capabilities that they need with the possibility to start small and scale up as the need for analytics increases. This is particularly critical for fast-growing companies that must align the cost and capabilities of software investments with the rate of growth and expansion of the operation (Ogbuokiri, Udanor, & Agu, 2015).

SMEs are adopting BI&A for various purposes (e.g., automation, having full control over their data, and enhancing their decision-making processes), but they only use the simple analytics functionality of BI&A. The most important perceived BI&A impacts for SMEs are related to obtaining business and customer insights, cost reductions, and competitive advantages (Llave, Hustad, & Olsen, 2018). Customer insight can increase sales and improve customer retention, and BI&A enables companies to create an intelligent campaign management that uses its customer data to select target groups for upselling and cross-selling (Llave, Hustad, & Olsen, 2018). For instance, big data analytics allow companies to predict customers' propensity to buy a certain product, to offer discounts and make personalized recommendations for future purchases, to predict or identify and fix potential failures of products, and to analyze online customer reviews and understand customers' experience with products or services, among other insights (Grover et al., 2018). In the next section, we expose the Mexican SMEs' context to understand how feasible it could be for these organizations to benefit from social media analytics to discover insights about the market that can ultimately be translated into actionable strategies.

2.3 SMALL AND MEDIUM ENTERPRISES IN MEXICO

In many economies of the world, SMEs play an active role in the economic development of the country. In Mexico, the National Institute of Statistics, Geography and Information (INEGI) in its last economic census reports that micro, small, and medium enterprises contribute 52% of the gross domestic product, with a workforce of 72% mainly distributed in the following activities: services (47%), commerce (25%), and manufacturing (28%) (INEGI, 2014). This indicates that SMEs are the support for development and industrial growth in Mexico. However, 52.5% of SMEs in Mexico tend to last less than two years, mainly due to the inability to be profitable.

To strengthen this type of company, in December 2002, the Law for the Development of the Competitiveness of Micro, Small, and Medium Enterprises was enacted in Mexico. This law defines competitiveness as the ability to maintain and strengthen profitability and participation in markets. Competitiveness lies in understanding the environment to generate competitive advantages and the implementation of strategies based on the resources and capabilities that distinguish the organization to maintain these advantages.

The truth is that, with an increasingly competitive environment, strengthening profitability and participation in the markets through an operational efficiency strategy is no longer sufficient. Unfortunately, in the current business environment, there are still many companies that follow this type of strategy, which can be dangerous for their long-term permanence. In its last Economic Census, the INEGI indicates that two of every three Mexican companies fail before or at the end of their first five years of life, which is strongly correlated with the size of the company. That is, when the business is smaller, the likelier it is that the business will fail in the first years of life, as four of ten microenterprises shut down during the first year.

The crucial question for SMEs is that, in dynamic environments characterized by increased competition and rapid technological change, a competitive advantage requires more than resources that are difficult to emulate. It seems that in this digital era, the degree of innovation and digitization varies significantly across sectors and businesses, which reinforce the dual economy. Sectors such as information and communications technology, finance, and the media are among the most digitized, while large traditional sectors such as retail, education, and agriculture are lagging behind. Smaller firms lack access to multiple enablers for technology adoption, such as finance, talent, a digital infrastructure, and often even information on relevant use cases. To compete effectively, SMEs must focus on the unique development of their dynamic capabilities, which should impede their imitation by their competitors. A dynamic capacity is not created simply by gathering a set of distinctive resources; its creation requires complex patterns of coordination between people and other resources. Considering the flexibility, simplicity, and low cost that social media tools could provide (McCann & Barlow, 2015; Ogbuokiri, Udanor, & Agu, 2015), SMEs can benefit from customer information and insights to improve and differentiate their services and products in order to compete and succeed.

In fact, there is a momentum by utilizing and building digital platforms, marketplaces, and intensify social media tools. For instance, the smartphone adoption in Latin America rose from 5% of connections in 2010 to 47% in 2015. Specifically, for Mexico in 2018 this percentage was 54.7% and mobile networks reached more than 42% (INEGI, 2018). Once they are online, Latin Americans will be among the world's most avid users of social networks—a development that gives companies effective ways to reach customers at scale (McKinsey, 2016b). In this regard, we propose that social media analytics tools could support and enhance Mexicans SMEs customer engagement as the first step to design and formalize a BI strategy.

2.4 SOCIAL MEDIA ANALYTICS TOOLS FOR CUSTOMER ENGAGEMENT IN MEXICAN SMEs

Social media represents a fundamental change of how information is being produced, transferred, and consumed. People around the world are creating content, which is also known as user-generated content, that may serve as a connection between organizations and consumers. Particularly, the data found in social media are so rich that they can be viewed as a dynamic evidence of human behavior that can provide opportunities to understand individuals, groups, and society in general (Batrinca & Treleaven, 2015). Therefore, companies and especially SMEs can benefit from tracking information on social media to gain feedback and insights in how to better serve consumers and succeed in this competitive market (Leskovec, 2011).

Large companies have the advantage of using technology and big data to enhance product and service quality, improve their operations, and establish stronger customer relationships. In contrast, it is common that most owners of small businesses follow their personal intuition and use their abilities to compete on the market and to face larger competitors (Ogbuokiri, Udanor, & Agu, 2015). This represents an important threat to small companies, which usually are struggling to survive in a competitive market. Considering that most big data tools are costly and complex to implement, and that most of them were built for larger organizations, SMEs have no chance of entering this trend. Fortunately, as technology evolves with the rise of information and communication technologies, SMEs now have the ability to use other available digital tools as a source of business analytics.

One of these tools is social media, which are web-based applications such as Instagram, Facebook, Twitter, and YouTube that allow for information dissemination, content generation, and interactive communications (Zeng et al., 2010). Furthermore, in addition to the popular platforms mentioned above, the term social media also refers to other sites such as blogs, wikis, and news (Batrinca & Treleaven, 2015). As the number of social media users continues to increase, the need for businesses to monitor and utilize these sites to their benefit also increases (Fan & Gordon, 2014).

Particularly for SMEs, social media represents a rich source of information and has the potential of becoming the next-generation BI platform. Considering that users share an incredible amount of information about themselves and that most platforms are free of charge to use, these businesses do not need to invest large amounts of money to gain important benefits. Therefore, the use of social media analytics with a large online community enables SMEs to better reach their customers at an unprecedented scale and lower costs (Kurniawati, Shanks, & Bekmamedova, 2013). Accordingly, for SMEs, the use of social media can represent an opportunity to gain information about the market, to establish a two-way communication with customers, and to improve customer engagement.

There is a widespread adoption of popular social media platforms by users around the world, and social media is already a critical part of the information ecosystem. These platforms have an unprecedented reach to users, consumers, voters, businesses, governments, and nonprofit organizations (Zeng et al., 2010). In this regard, social media analytics are now considered to be a discipline that helps companies measure, assess, and explain the performance of social media initiatives in the context of specific business objectives (Ruhi, 2014).

In recent years, Mexican SMEs' interest in using digital technologies has been increasing. There are many benefits that technology and social media in particular bring to SMEs; however, in Mexico, there is still a small percentage of companies that have decided to adopt them. Perhaps one of the motives that prevent the adoption of technology by Mexican SMEs is a lack of financing and knowledge (Forbes, n.d.). With respect to financing, as mentioned before, currently, there are many available tools and software that are free or accessible so that SMEs can engage in social media analytics without a large investment. Regarding knowledge, increasingly more programs target SMEs to help them adopt new technologies. In fact, Facebook Mexico has launched a free annual program in 2016 where more than 10,000 Mexican SMEs were trained to use popular social media platforms such as Facebook, Instagram, and WhatsApp for their businesses. In a competitive market where 75% of SMEs do not survive in the first two years, the main objective of this program is to reduce the technological gap in this key economic sector of the country and to help SMEs increase their success rate (Entrepreneur, 2018).

2.5 DIGITAL RECOMMENDATION FOR SMEs: A FRAMEWORK FOR CUSTOMER ANALYTICS ON SOCIAL MEDIA

The social media analytics process is now being addressed in recent literature; however, it appears that there is no clear consensus about the steps and activities that encompass this process. For example, one study proposes that the social media analytics process consists of five steps, which are to collect, monitor, analyze, summarize, and visualize social media data (Zeng et al., 2010). Furthermore, other literature suggests three phases, which are to capture, understand, and present. Both frameworks have several similarities that can be grouped into a single framework; however, they still lack the last step to truly utilize insights and turn them into actionable strategies. For this reason, this chapter integrates previous studies and proposes a new framework (Figure 2.1) with the four phases of listening, analyzing, insights, and strategy. These phases will be described next.

FIGURE 2.1 Proposed framework for customer analytics on social media. (Source: From own elaboration based on Fan & Gordon, 2014; Ogbuokiri, Udanor, & Agu, 2015; Zeng, Chen, & Li, 2010).

The first step in the process consists of monitoring the conversations that occur on different social media platforms related to the company, its products and services, the competition, and the industry. This listening phase requires a constant tracking and collection of what people are posting online that might be related to the company. Information about businesses, users, events, user comments and feedback, reviews, and other posts are extracted in this phase (Fan & Gordon, 2016). Considering that most SMEs have at least one social media platform and actively maintain it as a communication channel with customers, this monitoring phase is the first natural and logical step. However, despite SMEs' active use of social media, their activities are normally limited to sending messages to customers, posting updates about products and services, delivering promotions, discounts and giveaways, and even some monitoring regarding what people say about the company. Nevertheless, these activities are very basic. As mentioned before, with the explosion in the number of social media sites and the volume of use on them, these basic activities are not nearly enough (Fan & Gordon, 2016). Accordingly, most SMEs stay in the first stage of the social media analytics process. Presently, tracking online information about a particular topic is not a complicated process, and SMEs do not need to make a large investment for this purpose, as many tools are now available. Sources for this phase include the social media data in social networks, wikis, blogs, news, RSS, etc. For example, a company can track hashtags and, use photo recognition software, social media application programming interfaces (APIs) from Wikipedia, Twitter, and Facebook, and other software such as Google Trends, SocialMention, SocialSeeking, Gnip, Amplified Analytics, Lithium Social Media Monitoring, and Trackur, among others (Batrinca & Treleaven, 2015). The ultimate goal of this phase is to collect an important data source of information that will later be analyzed to obtain insights.

The second step in the process consists of analyzing the data collected in the listening phase. There are several tools that can be used for this purpose, from traditional statistical methods to other techniques such as text processing, sentiment analysis, and network analysis. Examples of the software that can be used in this phase include Google Fusion Tables, Zoho Reports, Tableau Public, Gephi, and NodeXL, among others (Batrinca & Treleaven, 2015). The type of analysis will depend on the business objective that the company has. For example, social media data can be used for a sentiment analysis in an attempt to understand the sentiment polarity (positive, negative, or neutral) or overall attitudes toward a particular topic, brand, product, or issue. Furthermore, data can be used to obtain emerging trends in the market and how they change over time. Companies can also use social media data to identify the key people or communities that have a significant voice in or contribution to a particular topic. Additionally, a competitive analysis can be conducted by tracking the posts made by the competitors, or the comments of consumers who talk about the competition's products or services. Finally, another type of analysis that a company can engage in with social media data is to monitor the comments related to past or current marketing campaigns to evaluate their performance (Kurniawati, Shanks, & Bekmamedova, 2013). Overall, the findings of this phase are crucial to move forward in the process of social media analytics, as the results will dictate the insights and strategies of the organization.

Next, the insight phase is responsible for making sense of the findings from the analyzing phase. This step requires seeking patterns and making conclusions about the reasons and meanings behind the conversations and posts found on social media. Most software used in the analyzing phase have an option to visualize information in a way that can be useful to draw insights at this moment. For example, there are some content analysis and text mining software such as ATLAS.ti MAXQDA and Nvivo, which are useful particularly to visualize data, find categories, and lead to insights about each emerging topic.

Finally, the next step known as strategy addresses the link between social media data and their metrics to strategic decisions and performance. Particularly, strategy aims to derive actionable information from social media to develop corresponding decision-making frameworks (Fan & Gordon, 2016). For example, a small retailer can use social media conversations and buying trends to take advantage of emerging sales opportunities. Furthermore, there is an opportunity to analyze the visitors of a website and track how they surf from page to page. This will allow understanding what strategies are engaging to users, detecting opportunity areas, and improving their users' experience with the brand (Ogbuokiri, Udanor, & Agu, 2015). These are examples of what social media analytics can result in.

Additionally, measurement of social media uses and valuation turns relevant to prove its benefits. For instance, McCann and Barlow (2015) expose ROI (return on investment) of social media can include various measures such as qualitative or quantitative data, informal or formal methods, and tangible or intangible benefits. The enormity and complexity of the measurement exercise may therefore be beyond the resources of many SMEs, being to have full evidence or be fully aware of the extent (if any) of the benefit.

Overall, it is important that SMEs do not underestimate the potential that social media can have for their business. Facebook in particular has been ranked as the number one social media tool that generates value for Mexican SMEs (González, 2018). SME managers/owners seeking to augment their marketing efforts on social media, multiple platforms should be employed since each media may offer unique benefits during their usage effect of combining the social media provides enhanced performance benefits (Odoom, Anning-Dorson, & Acheampong, 2017).

Social media data presents many opportunities and challenges to companies, and SMEs are no exception. It is true that although there is a vast quantity of online data available, the real challenge is to be able to analyze these data to gain meaningful insights and translate them into actionable strategies (Leskovec, 2011). This process requires companies to follow the above-mentioned methodology, but it also requires creativity to track the data from the best sources, to ask the right questions, and to discover insights with answers that can be translated into innovations and business opportunities. Despite these challenges, the message here is that big data is not only for large businesses but also for SMEs (Ogbuokiri, Udanor, & Agu, 2015).

The challenge for Mexican SMEs is how these companies manage and take advantage of such platforms, which may represent part of their competitive advantage. The ever-changing and complex social media environment requires organizations to plan their social media investments with care (McCann & Barlow, 2015).

2.6 CHALLENGES AND OPPORTUNITIES

In today's world, rapid advances and the growing adoption of disruptive technologies will create a new wave of opportunity to reinvigorate the small- and medium-sized firms (McKinsey, 2019). Digital platforms can help small businesses leapfrog and scale by increasing productivity and expanding to new markets through digital platforms, advanced analytics, and the Internet of Things.

The use of big data is not limited to large organizations, and SMEs can benefit from the enormous amount of data to make rapid and accurate decisions that improve their business functionality. This chapter recommends that SMEs seriously think about big data adoption to improve their business processes and obtain a better understanding of their consumers to develop strategies that improve customer engagement. Particularly, social media represents a rich source of information and has the potential of becoming the next-generation BI platform.

Companies can benefit from social media for both a rich source of information and a business execution platform not only for product design and innovation but also for customer engagement. For this purpose, this chapter addresses a social media analytics framework that can serve as a basis for SMEs to listen to and analyze online data, and come up with insights about the market that can ultimately be translated into actionable strategies. Overall, social media analytics are expected to contribute to the improvement of marketing strategies, create better customer engagement derived from a better service, improve brand awareness, serve as a basis for product development and improvement, and identify new business opportunities (Kurniawati, Shanks, & Bekmamedova, 2013).

Furthermore, this chapter identifies that the relevance of SMEs in Mexico and the difficulty of survival in an increasingly demanding market emphasize two essential aspects to boost their competitiveness. First, SMEs should incorporate digital strategies that allow them not only to stay in the market but also to grow and survive in the long term. They must seek to differentiate themselves continuously by creating competitive advantages through dynamic adaptation capacities with a strong emphasis on innovation strategies. This is only possible through a rigorous, and simultaneously flexible and dynamic strategic planning process. Second, public and private institutions should provide specialized consulting tools not only in key processes and activities of administrative management but also in digital, automatization, and data analytics that generate value that in the long term will allow SMEs to remain in the market. These institutions should also identify the distinctive resources and capabilities that they offer, so that these resources and capabilities can be transformed, renewed, and effectively incorporated into business strategies.

At present, the competitiveness of any company goes beyond improving profitability, implies dynamism to anticipate or timely face the constant changes in the environment, and indicates the incessant search for the company's survival. To be clear, digital is not a cure-all. However, if they are well harnessed, digital technologies could be powerful allies in the increasingly urgent struggle to restore national competitiveness.

A key success factor is therefore to examine customer analytics holistically, including IT, analytics, and execution/organizational setup, and to pragmatically improve on all dimensions. Further organizational setups that increase the value contribution of customer analytics are a management board that discusses customer-related topics and customer boards that fully embrace the customers' perspective (McKinsey, 2016a).

From a regional perspective, Latin American societies will need to work together to transform digital potential into value creation and to ensure that the digital value benefits different members of society, especially small businesses and vulnerable workers. Countries will need to address important challenges and risks, especially those relating to how automation will change occupations and the skills required to perform them. New technologies may also bring the risk of increased concentration for the most successful firms.

REFERENCES

Batrinca, B., and Treleaven, P. C. (2015), Social media analytics: A survey of techniques, tools and platforms. *Ai & Society*, 30(1), 89–116.

Chiang, R. H., Grover, V., Liang, T. P., and Zhang, D. (2018), Strategic value of big data and business analytics. *Journal of Management Information Systems*, 35(2), 383–387.

Entrepreneur. (2018), Facebook le echa la mano a las pymes mexicanas para crecer con su programa Impulsa tu empresa. Retrieved from https://www.entrepreneur.com/article/324717

Fan, W., and Gordon, M. D. (2014), The power of social media analytics. *Communication of the ACM*, 57(6), 74–81.

Fernandez, J. L., Gutierrez, J. E., Castro, L. A., and Rodriguez, L. F. (2016), Integrating business analytics into SMEs in Mexico: Challenges and opportunities. *Avances en Interacción Humano-Computadora*, 1(1), 41–45.

Forbes. (n.d.), Pymes mexicanas: llegó el momento de apostar por la tecnología. Forbes México. Retrieved from https://www.forbes.com.mx/brand-voice/pymes-mexicanas-llego-el-momento-de-apostar-por-la-tecnologia/

González, F. (2018), Así utilizan Facebook las PyMes mexicanas. Merca 2.0. Retrieved from https://www.merca20.com/asi-utilizan-facebook-las-pymes-mexicanas/

Grover, V., Chiang, R. H., Liang, T. P., and Zhang, D. (2018), Creating strategic business value from big data analytics: A research framework. *Journal of Management Information Systems*, 35(2), 388–423.

INEGI. (2014), Economic Census (2014). Retrieved from https://www.inegi.org.mx/contenidos/programas/ce/2014/doc/frrdf_ce2014.pdf

INEGI. (2018), Encuesta Nacional sobre Disponibilidad y Uso de Tecnologías de la Información en los Hogares 2018. Retrieved from https://www.inegi.org.mx/programas/dutih/2018/default.html#Tabulados

Iqbal, M., Kazmi, S. H. A., Manzoor, A., Soomrani, A. R., Butt, S. H., and Shaikh, K. A. (2018), A study of big data for business growth in SMEs: Opportunities & challenges. In 2018 *International conference on computing, mathematics and engineering technologies (iCoMET)* (pp. 1–7). IEEE.

Kurniawati, K., Shanks, G. G., and Bekmamedova, N. (2013), The Business Impact of Social Media Analytics. In proceedings of the 21st ECIS, 13, 13. Association for Information Systems.

Leskovec, J. (2011), Social media analytics: Tracking, modeling and predicting the flow of information through networks. In *Proceedings of the 20th international conference companion on World Wide Web.* (pp. 277–278). ACM.

Llave, M. R., Hustad, E., and Olsen, D. H. (2018), Creating value from business intelligence and analytics in SMEs: Insights from experts. Twenty-fourth Americas conference on information systems, New Orleans. Retrieved from https://www.researchgate.net/profile/Marilex_Llave/publication/330765646_Creating_Value_from_Business_Intelligence_and_Analytics_in_SMEs_Insights_from_Experts/links/5c5381d292851c22a39f5637/Creating-Value-from-Business-Intelligence-and-Analytics-in-SMEs-Insights-from-Experts.pdf

McCann, M., and Barlow, A. (2015), Use and measurement of social media for SMEs. *Journal of Small Business and Enterprise Development*, 22(2), 273–287.

McKinsey. (2016a, May), *Why customer analytics matter.* New York: McKinsey Global Institute.

McKinsey. (2016b, June), *Can Latin America reignite growth by connecting with consumers?* New York: McKinsey Global Institute.

McKinsey Report. (2019, May), *Latin America's missing middle: Rebooting inclusive growth.* New York: McKinsey Global Institute.

Odoom, R., Anning-Dorson, T., and Acheampong, G. (2017), Antecedents of social media usage and performance benefits in small- and medium-sized enterprises (SMEs). *Journal of Enterprise Information Management*, 30(3), 383–399.

Ogbuokiri, B. O., Udanor, C. N., and Agu, M. N. (2015), Implementing big data analytics for small and medium enterprise (SME) regional growth. *IOSR Journal of Computer Engineering*, 17(6), 35–43.

Raj, R., Wong, S. H. S., and Beaumont, A. J. (2016), Business intelligence solution for an SME: A case study. In *KMIS* (pp. 41–50).

Ruhi, U. (2014), Social media analytics as a business intelligence practice: Current landscape & future prospects. *Journal of Internet Social Networking & Virtual Communities*, 2014. DOI: 10.5171/2014.920553.

Wu, X., Zhu, X., Wu, G. Q., and Ding, W. (2013), Data mining with big data. *IEEE Transactions on Knowledge and Data Engineering*, 26(1), 97–107.

Zeng, D., Chen, H., Lusch, R., and Li, S. H. (2010), Social media analytics and intelligence. *IEEE Intelligent Systems*, 25(6), 13–16.

3 Data Management Software Solutions for Business Sustainability – An Overview

Mihaela-Filofteia Tutunea
Faculty of Business, Babeş-Bolyai University
Cluj-Napoca, Romania

CONTENTS

3.1 INTRODUCTION

In a globalized business environment where companies have adapted their activities to the digital nano-market and to the new rules imposed by it, in recent years, decision makers have increasingly begun to perceive that the most valuable assets for a company that wants sustainable development, are business data and professional human resource.

In recent years, aspects regarding the importance of business data, their integration at intra- and inter-organizational levels, and especially how to use them as key decision support for any company, have generated new business concepts such as Big Data, data governance (DG), data management (DM), and master data management (MDM); these concepts have also led to the development of new categories of software solutions dedicated to companies that provide them with competitive advantages and business sustainability.

IDC, in its "Data Age 2025" report, identified some extremely important issues related to data as a general concept, and business data as a specific concept; In this report, starting from the place, role, and evolution of these data, a new concept "global datasphere" is introduced as a vision of "digitization of the world" and wich is expected to grow from 33 Zettabytes (ZB) in 2018 to 175 ZB by 2025; they were also defined as "three primary locations where digitization is happening and where digital content is created: the core (traditional and cloud datacenters), the edge (enterprise-hardened infrastructure like cell towers and branch o ces), and the endpoints (PCs, smart phones, and IoT devices)" (Reinsel et al., 2018).

Gartner, Inc. (2019), in IT dictionary, has defined another concept, namely, Big Data, as being "high-volume, high-velocity and/or high-variety information assets that demand cost-effective, innovative forms of information processing that enable enhanced insight, decision making, and process automation." Also, it was indicated that Big Data has the following three important features: "data are numerous, data cannot be categorized into regular relational databases, and data are generated, captured, and processed rapidly" (Hashem et al., 2015, p. 99); for the Big Data, there were also identified five classification criteria, namely: "data sources, content format, data stores, data staging, and data processing" (Hashem et al., 2015, p. 100).

In the literature, delimitations have been made between DG and Big Data management; Big Data governance was defined as "a set of decision-making rights and responsibilities on data and information, executed in accordance with agreed processes, standards and models, describing who can take action, with what data and when, in accordance with predetermined methods and authorized access rights" (Kuiler, 2019, p. 1); also has been shown that Big Data management is focused on executing Big Data governance decisions, manages, coordinates, preserves, and protects Big Data resources (Kuiler, 2019, p. 1).

Master data, as a new concept, was defined as "the consistent and uniform set of identifiers and extended attributes that describes the core entities of the enterprise including customers, prospects, citizens, suppliers, sites, hierarchies and chart of accounts" (Gartner2, 2019); in the same context, MDM was introduced as "a technology-enabled discipline in which business and IT work together to ensure the uniformity, accuracy, stewardship, semantic consistency and accountability of the enterprise's official shared master data assets" (Gartner2, 2019).

MDM comprises "business applications, information management methods, and data management tools to implement the policies, procedures, and infrastructures that support the capture, integration, and subsequent shared use of accurate, timely, consistent, and complete master data" (Loshin, 2009, p. 8); an extremely important fact has also been highlighted, namely "master data management success depends on high-quality data" (Loshin, 2009, p. 17).

Regarding the adoption and implementation of MDM solutions at the level of the companies, it was shown that it is essential to consider the enterprise architecture and business model; in practice, have been identified as "three major types of initiatives based on its primary focus being operational or nonoperational data: analytical MDM, operational MDM, enterprise MDM" (Dalton and Mark, 2011, p. 10); Also, at the level of the companies, a set of "reasons for implementing the MDM solutions were identified, namely: cost reduction, risk management, and revenue growth" (Dalton and Mark, 2011, p. 24).

From the perspective of practitioners in business strategy, "data management provides a single view of data and, master data management provides a complete view of your organization's data" (CIO, 2019).

SAS Company, known as the leader in business analytics software and services, identified for any company "five data management best practices that support advanced analytics: simplify access to traditional and emerging data, strengthen the data scientist's arsenal with advanced analytics techniques, scrub data to build quality into existing processes, shape data using flexible manipulation techniques and share metadata across data management and analytics domains" (SAS, 2019).

SAP company, the market leader in enterprise application software, has pointed out that "data management should not be seen as a complex project to coordinate the various data management technologies, it being rather a continuous process of data integration, orchestration, governance, security, and other coordinated capabilities in a coherent manner through the global technology platform of the company" (SAP, 2019).

KPMG, a well-known global network of professional firms, highlights important issues related to the use and deployment of business intelligence (BI) and DM concepts and software solutions, starting from the importance of using BI and DM in DM generating competitive advantage; the same source, has also identified the methods used to implement these software solutions, as well as their core modules; thus BI and DM software solutions allow companies to define BI strategies and develop BI architecture tailored to the company's objectives; the next stage is the design and implementation of data warehouse solutions including operational data management (ETL [Extract Transform Load] tools); for any such software solution, it is obvious to ensure the quality of data used as well as DG, including KPI harmonization, MDM, and data clean-up; all these stages and modules allow for the generation of BI solutions, which become real support in monitoring and optimizing business processes and resources and substantiating decision-making in companies (KPMG, 2019).

From a statistical perspective, the global market of DM solutions is expected to reach $1.28,500.00 million by 2022, with an increase in CAGR (compound annual growth rate) of 11.0% from 2015 to 2022; and among the great players on this market are found the following: SAP SE; Intel Security; Teradata Corporation; SAS Institute, Inc.; Accenture; Talend; Oracle Corporation; IBM Corporation; Symantec Corporation; Informatica Corporation; GoldenSource Corporation; Cambridge Semantics, Inc.; Innovative Systems, Inc.; Phasic Systems, Inc.; and Solix Technologies Inc. (Reuters, 2019).

From another perspective, at company level, sustainable development analysis has generated a set of concepts such as "sustainability management, corporate sustainability, sustainability innovation and sustainable entrepreneurship and social business" (Schaltegger et al., 2016, p. 4). An aspect related to this was also identified, "business sustainability typology may be applied to organizations by considering the various dimensions of ownership, governance, strategy, and culture, thus providing an organizational roadmap toward business sustainability" (Thomas and Katrin, 2015, p. 16).

From a practical perspective, sustainable business development "involves the application of sustainability principles to business operations and can mean a variety of things – ecological sustainability, social sustainability or even sustained economic growth" (Sustainablescale, 2019).

By the synonym of business sustainability and corporate sustainability, this is defined as "the management and coordination of environmental, social and financial demands and concerns to ensure responsible, ethical and ongoing success" (Techtarget, 2019).

At the same time, from business perspective, business sustainability can be considered "as the ability of firms to respond to their short-term financial needs without compromising their (or others') ability to meet their future needs" (Bansal and DesJardine, 2014, p. 71).

Starting from the outlined picture, an analysis was made of solutions and software platforms in the DG, DM), and MDM category dedicated to companies, especially small and medium enterprises, and also a study on the use of these solutions by companies in Romania.

We consider that the results obtained can be a starting point for decision makers in companies to adopt these types of solutions and platforms, thus contributing to the use of business data as an essential tool in generating a massive competitive advantage and business sustainability.

3.2 MATERIALS AND METHODS

In order to get a complete picture of the solutions/platforms for DM and their use, the study included two phases: one dedicated to identifying software solutions in the categories of DG, DM, and MDM on the software market at the time of the study and the other dedicated to Romanian companies as users of these solutions.

In the analysis of software solutions, the DG, DM, and MDM software solutions were considered dedicated to small- and medium-sized companies and with financial plans based on demand or on monthly subscription, which from the financial perspective represents an important advantage for SMBs (small and medium-sized business); for this analysis, were also used sources of secondary information provided by specialized resources (www.getapp.com, https://sourceforge.net/, www.getapp.com, www.gartner.com/, www.datamation.com/, www.softwareadvice.com, etc.), statistical data, and web observation of the websites of producers/suppliers of these solutions; in the analysis was considered a mix of important criteria, namely, number of users, deployment model (web-based, cloud, installed, mobile), most reviewed solutions, the highest rated solutions, and list of features offered; we mention that all of the solutions listed offer free trial versions and cover all deployment models including mobile ones.

In the case of companies, it was used as a tool for identification and selection, the monitoring website Facebook Romania, facebrands.ro; the main categories were selected: companies (159), business (1545), hotels (325), IT&C (697), online stores (2677), Products (326), Health & Personal Care (3134), Web Services (548), Tourism & Vacations (1216); from each category 10% of the companies were selected, considering the number of fans as well as their evolution in time; the final sample consisted of 1060 companies; mobile-survey was used based on a questionnaire consisting of seven questions; the following issues were followed:

- identifying the perception of the importance of business data as decision support in the company;
- identifying the level of use of DM software solutions and their specific features actually used by companies;
- identifying the benefits and disadvantages perceived in the use of these solutions; and
- perception of companies on the need to allocate a larger budget for adopting/updating DM or MDM solutions in the near future.

The study was conducted between September and December 2018; only 215 valid questionnaires were obtained, thus achieving a relatively low response rate (20.28%).

3.3 RESULT AND DISCUSSIONS

3.3.1 DG, DM, and MDM Software Solutions

In the analysis, the identification of software solutions dedicated especially to small- and medium-sized companies was sought; for this reason, were not considered the major software vendors such as AWS (Amazon Web Services), IBM, Oracle, SAP, SAS, TerraData, Dell Boomi, etc., providing complex solutions dedicated to all categories of companies, regardless of their size, but with less affordable prices for SMBs; it is known that software vendors have very diverse financial and distribution plans, starting from the most preferred, tiered model, from basic to very complex features, then pay-per-user, also useful for SMBs, who usually avoid the one-time license plan, which involves a high cost of software acquisition.

In this context, it should be specified that for this category of software, there are extremely few open-source solutions; thus, in this class, we can only mention a few supplier companies:

- Pimcore (https://pimcore.com) that develops Pimcore Platform, covering DM and MDM solutions.
- Talend (https://www.talend.com), care ofera pachetul Open Studio for MDM.
- Best of BI (http://www.bestofbi.com), SQL Power DQguru – Data Cleansing and MDM Tool.
- LogiCoy (https://logicoy.com), provider for LogiCoy MDM.

It should also be underlined that open-source software solutions are viable and effective only for companies that have IT professional staff.

The analysis of software solutions in the category DG, DM and MDM, has led to the following results:

- *For the data governance solutions*, the following specific sets of features were considered: Access Control, Data Discovery, Data Mapping, Data

Profiling, Deletion Management, Email Management, Policy Management, Process Management, Roles Management, Storage Management; the results obtained are included in Table 3.1.

- *For data management solutions*, the following features were considered: Customer Data, Data Analysis, Data Capture, Data Integration, Data Migration, Data Quality Control, Data Security, Information Governance, MDM, Match & Merge, etc.; the results obtained are included in Table 3.2.
- *For the MDM solutions*, the following features were considered:

DG, Data Masking, ata Source Integrations, Hierarchy Management, Match & Merge, Metadata Management, Multi-Domain, Process Management, Relationship Mapping, Visualization; the results obtained are included in Table 3.3.

The results obtained and presented in tabular form allow the visualization of the fact that the software market presents an important, varied and complex offer of software solutions in the DM category; As a result, any decision maker in companies, including small and medium-sized businesses, after careful consideration of the features offered by these software solutions, can choose to adopt the most useful and efficient solution for their own company.

TABLE 3.1
Data Governance Software Solutions

Product	Provider, Website
Alfresco Content Services	Alfresco, www.alfresco.com
Big Data	Informatica, www.informatica.com
Cloudera Navigator	Cloudera, www.cloudera.com/
Collibra	Collibra, www.collibra.com
Commvault Activate	Commvault, www.commvault.com
Data Stream Manager	Insite Innovations and Properties, www.datastreams.io/
Data3Sixty	Infogix, www.infogix.com
Dattamza	Dattamza, dattamza.com
DEMS	Global Data Excellence, www.globaldataexcellence.com
Digital Insights	Namogoo, www.namogoo.com/
dspConduct	BackOffice Associates, www.boaweb.com/
erwin Data Governance	Erwin, www.erwin.com/
FileFacets	FileFacets, www.filefacets.com/
Global Relay Archive	Global Relay Communications, www.globalrelay.com/
Information Value Management	Datum, www.datumstrategy.com/
inSync	Druva, www.druva.com
OneSoft Connect	OneSoft Connect, onesft.com
Profisee	Profisee Group, www.profisee.com
Silktide Insites	Silktide, www.silktide.com
TrustMAPP	Secure Digital Solutions. www.trustmapp.com
WebAssurance	ObservePoint, www.observepoint.com/

TABLE 3.2
Data Management Software Solutions

Product	Provider, Website
Actian X	Actian, www.actian.com/
Actifio	Actifio, www.actifio.com/
Data Integration	Talend, www.talend.com
Data Management Platform	OnAudience, www.onaudience.com
Delphix	Delphix, www.delphix.com
Dharma	Dharmaplatform, https://dharmaplatform.com/
Indigo DQM	Indigo Scape DRS Data Reporting Systems, www.indigoscape.net
Information Value Management	Datum, www.datumstrategy.com/
Intelligent Data Platform	Informatica, www.informatica.com
QlikView	Qlik, www.qlik.com
skyWATS	Virinco, https://www2.skywats.com/
Veera Construct	Rapid Insight, www.rapidinsightinc.com
Worksheet Systems	FalconSoft, worksheet.systems

TABLE 3.3
Master Data Management Software Solutions

Product	Provider, Website
2i	B&C Technologies, www.bnctech.com/homepage.html
Ataccama One	Ataccama, https://one.ataccama.com/
Delphix	Delphix, www.delphix.com
EBX	Tibco, www.orchestranetworks.com
Enterworks Enable	EnterWorks, www.enterworks.com
Informatica MDM	Informatica, www.informatica.com
Keen	Keen Labs, www.keen.io
Magnitude MDM	Magnitude Software, https://magnitude.com/
MaPS System (MDM-PIM-DAM)	MaPS System,www.maps-system.com/
MasterDataOnline	Prospecta Software, www.prospectasoftware.com/
MDM Platform	Talend, www.talend.com
MDMCenter	Riversand, www.riversand.com/
Omni-Gen Data Management	Information Builders, www.informationbuilders.com
Profisee	Profisee Group, www.profisee.com
STEP	StiboSystems, www.stibosystems.com/
TIBCO MDM	Tibco, www.orchestranetworks.com
Triniti Master Data Management	Triniti, www.triniti.com
Vin MDM	Vinculum Group, www.vinculumgroup.com

3.3.2 The Study Regarding the Use of Data Management Software Solutions by Romanian Companies

The analysis of the data gathered following the social media survey carried out allowed the identification of a set of aspects regarding

- the perception of the importance of business data as a decision support in the company, identified a segment of 48.37% of the respondents indicating that these are important, 28.34% of the companies find them very important, and 23.29% quite important; it should be underlined that none of the respondents considered these data to be a little or no important;
- identifying the type of data used mainly in decision support has brought to the fore customer data (62.35%) and product data (59.64%), followed by social media, e-mail, etc. (49.33%) (Figure 3.1).
- identifying the use of the indicated software solutions resulted in the delimitation of a very important segment of nonusers (67.87%).
- the identification of the software solutions' features mainly used by the users respondents indicated the use of customer data (58.19%) and data analysis (55.78%), followed by data migration (43.11%), MDM (23.01%); it is observable and worrying that data security is not part of the set of preferred features by the respondents (Figure 3.2).
- the list of benefits identified by the companies surveyed ranked first: cutting costs (58.23%), followed by increased revenue (46.11%), reducing losses (43.09%), better decision-making (38.56%), improving productivity

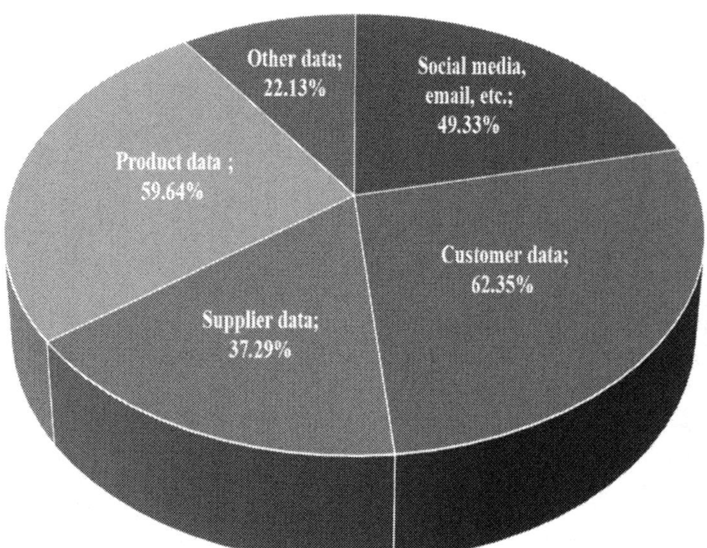

FIGURE 3.1 The type of data used in decision-making.

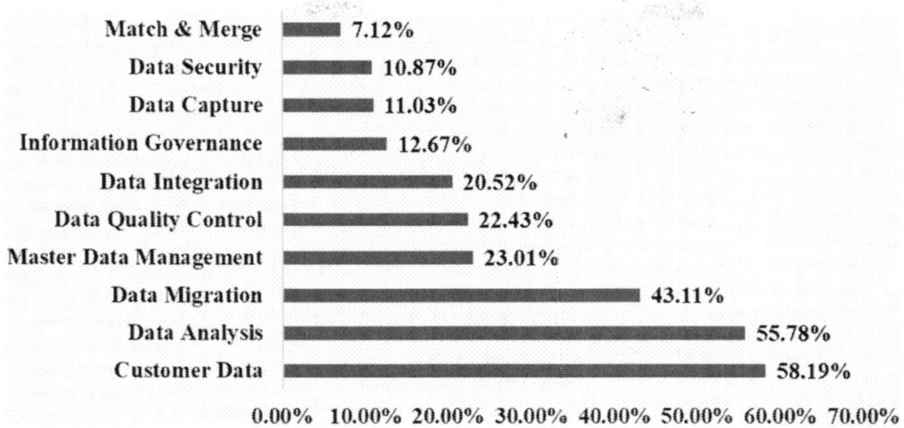

FIGURE 3.2 The software solutions' features mainly used. (Chart made by the author.)

(38.10%); the ease of application development is considered to be the least advantage of these solutions (Figure 3.3).

- the main disadvantages perceived by respondents are high costs (76.86%), difficulties in integrating software solutions (68.34%), inappropriate hardware infrastructure (56.73%), lack of IT/specialized staff (52.08%), data quality issues (43.23%); it is noticed that data security risk is not a concern or an important disadvantage for the respondents (Figure 3.4).
- the perception of the need to allocate a larger budget for the adoption/updating of DM or MDM platforms in the near future has formed a fairly low segment (56.78%), thus, forming a very important segment of users, which will continue to maintain the same outdated solutions or some with an insufficient set of features.

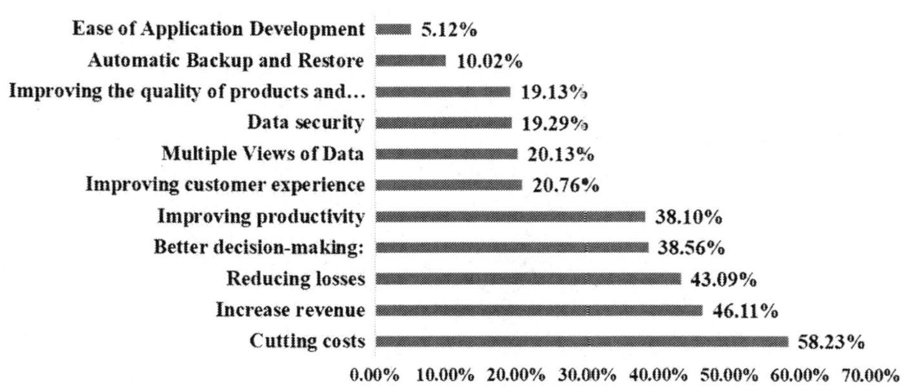

FIGURE 3.3 Benefits in using DM software solutions. (Chart made by author.)

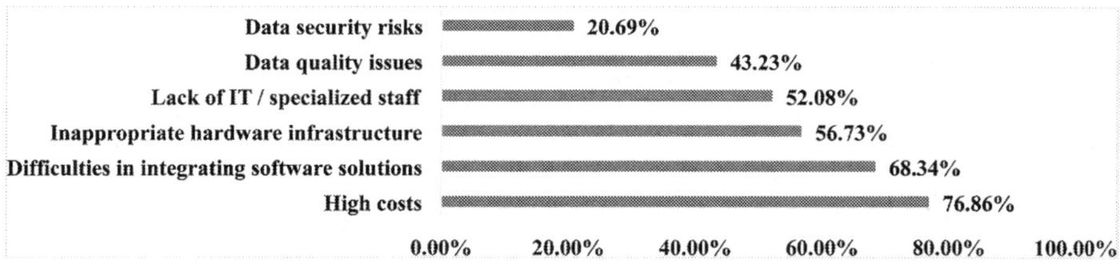

FIGURE 3.4 Disadvantages perceived in the use of DM software solutions.

3.4 CONCLUSIONS

Considering aspects of the importance of business data for the sustainable development of any company, as well as the rapid multiplication of data at least in terms of their type, origin, locations, and usage destinations, it can easily be noticed that the need to use DM software solutions at any company level has become a real fact.

The results obtained by analyzing software solutions from the specific DG, DM, and MDM categories available on the software market and dedicated to companies, indicate a rich and differentiated offer in the commercial software category that can cover both the acquisition needs and opportunities for SMBs; at the same time, the very low segment of open-source bidders for such solutions is visible; considering the relatively high acquisition costs of these solutions, which for SMBs can be a considerable financial effort, it may be helpful for providers of these solutions to reconsider their software packages both as a features list and as a financial plan to facilitate easier access for small businesses.

The study on the use of DM software solutions by a set of Romanian companies has allowed the outline of a local corporate picture, including companies that

- are aware of the importance of their business data, both as decision support and as a competitive advantage in a sustainable perspective;
- are oriented towards considering and using business data in domains as customers, products, and social media, but other types of data (master data) are not well targeted;
- forms a relatively small segment of DM software users and exploits only a minimal list of features of these solutions;
- perceive, on the one hand, cutting costs and increase revenue as the main advantages of using these solutions, and, on the other hand, high costs and difficulties in integrating software solutions as the main drawbacks in adopting and implementing the indicated solutions.

Based on the results obtained, we believe that both the software solution providers of the studied category and the companies as their potential users can identify a set of information that will allow them an advantageous and sustainable repositioning in a globalized and continually transforming market.

REFERENCES

Cervo Dalton, Allen Mark, 2011, Master Data Management in Practice. Achieving True Customer MDM, John Wiley & Sons, Inc., Hoboken, New Jersey, pp. 10–24.

CIO, 2019, The capabilities and roles of world-class, master data management, https://www.cio.com/article/3258822/the-capabilities-and-roles-of-world-class-master-data-management.html, accessed in February 2019.

Dyllick Thomas, Muff Katrin, 2015, Clarifying the Meaning of Sustainable Business: Introducing a Typology From Business-as-Usual to True Business Sustainability, Organization & Environment, pp. 1–19, https://www.bsl-lausanne.ch/wp-content/uploads/2015/04/Dyllick-Muff-Clarifying-Publ-Online.full_.pdf

Gartner, 2019, IT dictionary, https://www.gartner.com/it-glossary/big-data/, accessed in February 2019.

Gartner 2, 2019, Master Data Management, https://www.gartner.com/it-glossary/master-data-management-mdm/, accessed in January 2019.

Ibrahim Abaker Targio Hashem, Ibrar Yaqoob, Nor Badrul Anuar, Salimah Mokhtar, Gani Abdullah, Samee Ullah Khan, 2015, The rise of "big data" on cloud computing: Review and open research issues. Information Systems, 47, 98–115, https://umexpert.um.edu.my/file/publication/00001293_117865.pdf

KPMG, 2019, https://home.kpmg/ro/en/home/services/advisory/consulting/business-intelligence/bi-data-management.html, accessed in March 2019.

E. W. Kuiler, 2019, Data Governance. In: L. Schintler, C. McNeely (eds). Encyclopedia of Big Data. Springer, Cham, pp. 1, 4, https://link.springer.com/referenceworkentry/10.1007%2F978-3-319-32001-4_306-1

David Loshin, 2009, Master Data Management, Morgan Kaufmann OMG Press, Burlington, MA, pp. 8–17.

Pratima Bansal, Mark R. DesJardine, 2014, Business Sustainability: It is About Time, Strategic Organization, 12(1), 70–78. https://journals.sagepub.com/doi/pdf/10.1177/1476127013520265

David Reinsel, John Gantz, John Rydning, 2018, The Digitization of the World From Edge to Core, p. 3, https://www.seagate.com/www-content/our-story/trends/files/idc-seagate-dataage-whitepaper.pdf, accessed in march 2019.

Reuters, 2019, https://www.reuters.com/brandfeatures/venture-capital/article?id=50390, accessed in March 2019.

SAP, 2019, https://www.sap.com/cmp/dg/sme-data/typ.html#pdf-asset=4c24b773-197d-0010-87a3-c30de2ffd8ff&page=5, accessed in January 2019.

SAS, 2019, https://www.sas.com/content/dam/SAS/en_us/doc/whitepaper1/data-management-for-analytics-best-practices-107769.pdf, accessed in January 2019.

Stefan Schaltegger, Erik G. Hansen, Florian Lüdeke-Freund, 2016, Business Models for Sustainability: Origins, Present Research, and Future Avenues, Organization & Environment, 29(1), 3–10. https://journals.sagepub.com/doi/pdf/10.1177/1086026615599806

Sustainablescale, 2019, http://www.sustainablescale.org/AttractiveSolutions/Sustainable-BusinessPractices.aspx, accessed in March 2019.

Techtarget, 2019, http://whatis.techtarget.com/definition/business-sustainability, accessed in January 2019.

4 A Paradigm Shift in Accounting and Auditing in the *Era* of Big Data

Adelaide Martins
Research on Economics, Management and Information Technologies (REMIT), University Portucalense, Porto, Portugal

Ana Paula Silva
Polytechnic Institute of Viana do Castelo, Viana do Castelo, Portugal

Alexandra Fontes
Research on Economics, Management and Information Technologies (REMIT), Polytechnic Institute of Viana do Castelo, Viana do Castelo, Portugal

CONTENTS

4.1 INTRODUCTION

Business intelligence, analytics, and the related field of Big Data analytics have become increasingly crucial in both the academic and the corporate agenda of many organizations over the past two decades (Chen, Chiang, & Storey, 2012; Del Vecchio, Di Minin, Petruzzelli, Panniello, & Pirri, 2018). Business intelligence and analytics technologies facilitate data collection, analysis, and information delivery to support decision-making processes (Rikhardsson & Yigitbasioglu, 2018; Warren, Moffitt, & Byrnes, 2015). Given that accounting is a decision-supporting activity in providing relevant and reliable information (Ballou, Heitger, & Stoel, 2018), there is an evident connection among business intelligence, analytics, and accounting (Fay & Negangard, 2017; Rikhardsson & Yigitbasioglu, 2018). This is true regardless of accounting specialty – financial accounting, managerial accounting, tax advising, or auditing (Fay & Negangard, 2017; Schneider, Dai, Janvrin, Ajayi, & Raschke, 2015).

The data ecosystem is exponentially expanding, and this environment exhibits a dynamically changing set of characteristics (Enget, Saucedo, & Wright, 2017). Big Data is revolutionizing the business world as it provides companies and their external stakeholders with sharpened insights from digging into the vast volumes of data available nowadays, meaningfully informing and contributing to decision-making (Fay & Negangard, 2017; Griffin & Wright, 2015; Zhang, Yang, & Appelbaum, 2015). As put forward by Zhang et al. (2015), Big Data "pervades every sector and function of the global economy" (p. 469). In today's data-driven decision-making environment, the economy is being supported by real time processes. The increasing demand for predictive information requires skills to handle large datasets and to perform real time analyses (Appelbaum, Kogan, Vasarhelyi, & Yan, 2017; Richins, Stapleton, Stratopoulos, & Wong, 2017). It follows that Big Data is changing how accounting data is understood and reported (Griffin & Wright, 2015), providing a unique opportunity for accountants to capitalize on Big Data analytics by leveraging accounting data and its sound business competencies – thereby, conquering a redoubtable role in supporting effective decision-making (Lawson et al., 2014; Rikhardsson & Yigitbasioglu, 2018; Sledgianowski, Gomaa, & Tan, 2017). Like accountants, auditors are also expected to build up competencies in Big Data analytics in order to remain competitive, namely to achieve high levels of efficiency and improve financial statements audits, which shall no longer rely on sampling, but rather be sustained by entire populations of data (Cao, Chychyla, & Stewart, 2015; EY, 2018; Richins et al., 2017). Yet, the truth remains that many accountants and auditors, like business professionals in general, lack the fundamental skill set around data and analytics required to improve decision-making within a company, as well as to increase auditing effectiveness and efficiency (Enget et al., 2017; Fay & Negangard, 2017). In short, business intelligence, analytics, and Big Data present challenges and opportunities for accounting and auditing (Griffin & Wright, 2015).

It is no surprise that business education is beginning to adapt to the abysmal paradigm shift brought about by Big Data, business intelligence, and analytics as accounting educators are expected to integrate these topics into the curricula to meet the emerging needs of the accounting profession (Association to Advance Collegiate Schools of Business [AACSB], 2016; McKinney, Yoos, & Snead, 2017; Sledgianowski et al., 2017). A major concern is the slow pace at which this is taking place (Griffin & Wright, 2015; McKinney et al., 2017). The changes needed for accounting education to improve students' preparedness to effectively utilize Big Data are just beginning to emerge (McKinney et al., 2017).

Despite this current focus, understanding of business intelligence solutions and their impacts on decision-making and control in the accounting domain is very limited regardless of specialty (Fay & Negangard, 2017; Granlund, 2011; Rikhardsson & Yigitbasioglu, 2018). Our main purpose is to frame important issues for accounting and external auditing emerging from the *era* of Big Data. We expect this chapter will motivate a fruitful debate among students, academics, practitioners, and professional associations concerned, as to how such professions could best respond to the Big Data opportunities.

4.2 BUSINESS INTELLIGENCE, ANALYTICS AND BIG DATA

Business intelligence and analytics are often defined as the techniques, technologies, systems, practices, methodologies, and applications that analyze critical business data to help organizational decision makers to take better and informed decisions (Chen et al., 2012; Rikhardsson & Yigitbasioglu, 2018). It was not before the 1990s that business intelligence became a popular term in the business world (Chen et al., 2012). In the late 2000s, business analytics was introduced to represent the key analytical component in business intelligence (Chen et al., 2012; Davenport, 2006). Business intelligence and analytics are an "umbrella term" as they encompass a variety of technologies and methodologies that enable organizations to collect data from internal and external sources, prepare it for analysis, run queries, and produce reports, dashboards, and data visualizations in order to make results available to end users (Rikhardsson & Yigitbasioglu, 2018, p. 38). In recent years, Big Data and Big Data analytics have been used to describe the datasets and analytical techniques in applications that are so large and complex that traditional technology and information systems are inadequate to process and analyze them (Cao et al., 2015; Chen et al., 2012; Sledgianowski et al., 2017; Vasarhelyi, Kogan, & Tuttle, 2015; Warren et al., 2015). Big Data emerges from a technological environment where almost anything can be recorded, captured digitally or measured, and turned into data (Cao et al., 2015), providing an increased variety of sources and types of (voluminous) usable data (Janvrin & Watson, 2017). Big Data can be originated from structured sources, such as data from firms' transaction processing systems (e.g., point-of-sales systems, inventory management systems, customer/supplier relationship management systems – orders, sales revenues, receivables, payables, time sheets, product defect rates, personnel information), but also, and mostly, from unstructured sources (e.g., e-mails, audio streams, social media postings, news media, sensor recordings, surveillance videos, Internet click streams) (Appelbaum et al., 2017; Cao et al., 2015; Janvrin & Watson, 2017; Richins et al., 2017; Warren et al., 2015). The qualification of a particular dataset as Big Data is determined by whether or not it pushes the capability limits of the information systems working with such data (Enget et al., 2017; Vasarhelyi et al., 2015). It follows from the above that Big Data analytics can be described as a process of inspecting, cleaning, transforming, and modeling Big Data to extract useful insights and knowledge to assist management decision-making (Cao et al., 2015).

Big Data has been described through three Vs that make it seem overwhelming to work with: immense *volume* of the dataset, high *velocity* of data generation, big *variety* of data sources, and a fourth V is often suggested – (uncertain) *veracity* (Appelbaum et al., 2017; Cao et al., 2015) and even a fifth V – *value* (usefulness or cost-benefit) of data collection (Del Vecchio et al., 2018; Janvrin & Watson, 2017, p. 3, emphasis added; Sen, Ozturk, & Vayvay, 2016). Volume and velocity have been present since the 1990s when enterprise resource planning (ERP) systems have developed to bring expanded data storage power and better computational power (Appelbaum et al., 2017; Janvrin & Watson, 2017).

Big Data has plenty of uses in different fields. For example, it has been used to target customers in marketing, study voter demographics in political campaigning, evaluate teams and players in sports, identify threats in national security, identify crime suspects in law enforcement, and predict stock price averages (Cao et al., 2015). As far as businesses are concerned, Big Data analytics permits better insights and predictions of organizational stakeholders' behavior, thereby offering the potential to improve business processes and disrupt business models (Del Vecchio et al., 2018; Rikhardsson & Yigitbasioglu, 2018). Warren et al. (2015) suggest that Big Data analytics enhances companies' abilities to measure customer and employee satisfaction levels, as well as to evaluate and predict business performance. Organizations can now use browsing preferences, website clicks, video streams, environmental factors, and social media postings to forecast customers' purchasing patterns, improve customer satisfaction, and increase their business results (Fay & Negangard, 2017). Enget et al. (2017) strengthen that as organizations continue to expand the scope of their information systems from traditional data processes to extensive automated search methods, they increasingly benefit from a well-informed perspective leading to improved decision-making and, ultimately, to achieve the measured business outcomes they desire. In other words, companies that effectively and efficiently utilize Big Data have the potential to gain significant competitive advantages, including cost avoidance, increased profits, clear(er) thinking, and new product/service development (Enget et al., 2017).

ERP systems have been developed to allow more internal and external data of (non)financial and (un)structured nature to undergo data analytics techniques in order to describe what has happened (descriptive data analytics, such as ratio analysis, dashboards or pie charts, clustering models, text mining models, and descriptive statistics), predict what could happen (predictive data analytics, such as predictive and probability models, forecasts, statistical analysis, and scoring models), and prescribe what should be done (prescriptive data analytics based on mathematical simulation models or operational optimizations models) (Appelbaum et al., 2017). Predictive analytics requires the knowledge acquired from descriptive analytics, and prescriptive analytics draws from descriptive and predictive analytics results (Appelbaum et al., 2017). The most advanced business analytics are no longer based on statistical data analysis, but rather on machine learning, artificial intelligence, deep learning, text mining, and data mining (Appelbaum et al., 2017).

Rather than threatening the accounting and auditing professions, the unprecedented explosion of Big Data, through Big Data analytics, including accounting analytics, creates plenty of opportunities for improved accounting (EY, 2018; Fay & Negangard, 2017; Richins et al., 2017; Schneider et al., 2015) and for improved auditing (Cao et al., 2015; Earley, 2015; Fay & Negangard, 2017; Richins et al., 2017; Schneider et al., 2015; Yoon, Hoogduin, & Zhang, 2015) as further developed below.

4.3 THE OPPORTUNITIES OF BIG DATA ANALYTICS FOR THE ACCOUNTING AND AUDITING PROFESSIONS

The goal of accounting has always been to create and provide information to internal and external decision makers (Warren et al., 2015). The importance of reliable accounting data to assist businesses' stakeholders is widely acknowledged (e.g., EY,

2018; FASB, 1980; Maines & Wahlen, 2006; Melville, 2015). Inaccurate accounting records and financial reporting may result in suboptimal business decisions, loss of market share, loss of customers, financial instability, among other problems (EY, 2018). As early as 1961, Davidson and Trueblood argued "there are symptoms of management dissatisfaction with current accounting systems. Many managers believe the accounting function has failed to adjust its objectives and activities to the decision-making requirements of a changing business world" (p. 577). The authors already contended that "accounting for decision-making" (Davidson & Trueblood, 1961, p. 577) was a major challenge to the profession. The international accounting convergence phenomenon in place toward the International Financial Reporting Standards (IFRS) has brought forward the importance of accounting as an information system that shall be capable of providing useful and comparable information to internal and external stakeholders.

Accountants and auditors have traditionally focused on only two of the Vs that have been used to describe Big Data – veracity and value (Janvrin & Watson, 2017). For example, larger firms have long hired external auditors to provide assurance services to find out about veracity. Yet, the business competition level in a technological environment where industries 4.0 proliferate takes much more than backward looking financial statements. Indeed, it demands real time and predictive reporting sustained by more varied and voluminous (Big) data (Appelbaum et al., 2017; Richins et al., 2017). While historically business managers have been assisted by information generated from accounting records (Warren et al., 2015), it has been argued that there is a 94% likelihood that accounting and auditing jobs will eventually become automated (Frey & Osborne, 2013), which is already the case for most entry-level accounting and auditing tasks. However, it is difficult to replicate human key skills that Big Data analytics has the potential to leverage, thereby preventing these professions from facing extinction in favor of information technology professionals. Furthermore, the IFRS model, almost worldwide adopted, is principle-based, thereby reinforcing the relevance of accountants' professional judgment. Similarly, the International Standards on Auditing (ISAs) call for professional judgment to a large extent. For example, the sufficiency and appropriateness of the collected audit evidence is a matter of professional judgment.

As to accountants, it is becoming well established that their role and responsibility shall evolve from routine nature tasks culminating with the reporting of historical data toward becoming increasingly more advice-driven and consultative in nature, where key financial and operational drivers of shareholder value need to be identified, measured, and managed (Appelbaum et al., 2017; Richins et al., 2017). As argued by Rikhardsson and Yigitbasioglu (2018, p. 43), in today's data-driven environment, accountants might need to undertake new roles as their "monopoly" over reporting erodes. In the Big Data *era*, accountants must secure or else develop a strong expertise in strategy formulation and implementation in order to assist businesses in leveraging Big Data into implementable successful strategies, monitor the attainment of strategic objectives, as well as recommend the necessary corrective actions (Richins et al., 2017). By supplementing their sound business expertise with Big Data analytics skills, accountants can broaden their monitoring techniques to include unstructured data, thereby maximizing the potential to identify areas of improvement and opportunities (Richins et al., 2017). Warren et al (2015) provide a fairly comprehensive review on how unstructured data may improve financial information. Therefore, Big Data opportunities for accounting professionals emerge not only from transactions and balances that usually reside in a company's ERP (EY, 2018), but also from supplementary unstructured data that, after processing through analytics, may reveal matters of managers' interest (Cao et al., 2015; Richins et al., 2017; Warren et al., 2015). In this vein, the accounting professional may become a "forward thinking strategic partner in the organization" (Dzuranin, Jones, & Olvera, 2018, p. 24).

According to Richins et al. (2017), problem-driven analysis of unstructured data and exploratory analysis of both structured and unstructured data are becoming increasingly important in the Big Data world. Data scientists have a comparative advantage over accountants in terms of exploratory analysis of (un)structured data as this can be highly technical, leveraging tools from statistics, artificial intelligence, and database management (Richins et al., 2017). Given that accountants already excel at problem-driven analysis of large structured datasets (e.g., through DuPont Analysis in financial accounting or Balanced Scorecard in management accounting), they are well positioned to evolve and move to the realm of Big Data analytics by taking the lead in problem-driven analysis of unstructured data. As problems arise, accountants can leverage their business expertise by employing Big Data analytics on both structured and unstructured data (Richins et al., 2017). For example, accountants may: (i) perform content analysis of social media to evaluate public opinion about a firm's corporate social responsibility (Richins et al., 2017) or to predict product demand and returns (Fay & Negangard, 2017), (ii) use textual analysis of employee e-mails to detect and prevent fraud, (iii) assess company performance in many respects by combining financial statement data with information from tweets (Richins et al., 2017), or (iv) analyze customer phone calls and e-mails to identify troublesome product/service features and make recommendations accordingly, as well as improve estimates for warranty liability (Richins et al., 2017; Warren et al., 2015).

Even in terms of exploratory analysis, accountants are not replaceable by data scientists. As data volumes increase and more types of data are included in business intelligence and analytics systems, challenges regarding "accuracy, reliability, consistency, completeness, and verifiability emerge and new methods are needed to assess and improve data quality for decision-making" (Rikhardsson & Yigitbasioglu, 2018, p. 48). Big Data potentially contains overwhelming amounts of useless and/or unreliable data (Fay & Negangard, 2017) from which it emerges the opportunity for accountants to act in a support role to data scientists' exploratory analyses on Big Data. Their understanding of the business and its complex operations is crucial to sort out the relevant from the irrelevant content to include in exploratory analyses (both structured and unstructured), discard spurious correlations often found, ensure that correct inferences are made when letting the data speak for themselves, and recommend strategy formulation accordingly (Richins et al., 2017). In this way, accountants and auditors (Alles, 2015) are prime candidates to play a leading role in responding to the fundamental concern of firms engaging in Big Data analytics – which is determining whether underlying data is relevant and trustworthy since otherwise results could be dangerously flawed (Richins et al., 2017). Moreover, blindly following results from Big Data analytics, without sound business expertise, can be disastrous. As firms increasingly rely on Big Data analyses to direct their marketing, product development, and strategies, it becomes paramount to ensure validity of Big Data results (Alles, 2015). For example, a firm can use Big Data analytics to identify low-competition areas as potential areas for expansion. However, increased sales followed by declining inventories could either reflect difficulties in keeping up with demand due to exceptional growth, or rather, reflect liquidation by firms preparing to exit the market. In this case, the acumen of an accountant or an auditor (Alles, 2015) will be imperative to establish a difference between the two scenarios, namely by complementing analysis with unstructured data, such as textual analysis of competitors' management discussion (Richins et al., 2017).

Research on management techniques shows that they do not have changed significantly after the implementation of ERP systems. Besides, descriptive analytics are still the type of business analytics mostly utilized while predictive, and especially prescriptive analytics, remain underutilized (Appelbaum et al., 2017). For example, prescriptive accounting analytical tools are beyond the offer of EY, one of the Big Four accounting firms, but it offers descriptive accounting tools and predictive analytics based on statistics (Statistical Package for the Social Sciences [SPSS], Statistical Analysis System [SAS], and R) (EY, 2018). In a global survey of executives by KPMG, business complexities were identified as their greatest challenge by 94% of executives (Richins et al., 2017). Another survey including 2000 firms found that 86% were struggling to extract valuable information from Big Data (Fay & Negangard, 2017). In sum, evidence suggests that while many businesses invest significant resources to collect and process Big Data, most of them fail to reap and derive the expectedly sharpened insights and knowledge. They are still far from taking full advantage of all the functions of their enterprise systems, and this creates numerous potential opportunities for accountants to use analytical tools to go beyond providing descriptive reports, by conducting predictive and prescriptive analysis to help businesses to gain competitive advantage (Appelbaum et al., 2017). In this context, Rikhardsson and Yigitbasioglu (2018) contend that accountants are expected to play a more active role in assisting information technology personnel, namely by aligning the characteristics of the business intelligence analytics systems with the requirements and features of organizations inside users. Given their academic and professional preparedness, particularly their familiarity with financial information and firms' systems and structures, accountants are best positioned to help organizations derive valuable insights from globally integrated data capture and extraction tools – thereby assisting to achieve a deep business understanding convertible into successful strategies through full coverage of all the relevant transactions and risks. For instance, it has been suggested that the management accounting function might become integrated into a broader analytical function in the organization, together with customer analytics, process analytics, and environmental analytics (Rikhardsson & Yigitbasioglu, 2018). In this regard, as remarked by Rikhardsson and Yigitbasioglu (2018, p. 45), management accountants will "have to share responsibility for traditional management accounting analysis, such as profitability analysis for product and customer mix decisions, outsourcing decisions, capital budgeting valuations, stock optimization and incentive system design with other functions".

Like accountants, auditors have long used large structured datasets, typically data generated from clients' accounting information systems, with a problem-driven approach (Richins et al., 2017; Yoon et al., 2015). In the Big Data future, in order to remain competitive in the assurance services market, auditors, as much as accountants, must build up competencies in Big Data analytics, particularly in the problem-driven analysis of unstructured data and in the exploratory analysis of structured and unstructured data (Richins et al., 2017). Three of the Vs that describe Big Data – volume, velocity, and variety – contribute favorably to the "sufficiency" requirement of the audit evidence to be collected. The appropriateness (reliability) of the audit evidence may also be favored considering that Big Data is mostly externally generated and directly acquired by auditors, which, in turn, further benefits the sufficiency requirement (Yoon et al., 2015). In addition, reliability of audit evidence can be assessed more easily in a Big Data environment than in traditional manual audit inspection – for example, global positioning system data provides a more reliable data source than shipping documents to verify shipments (Yoon et al., 2015). Apart from providing an independent benchmark to evaluate internal audit evidence, Big Data may also be valuable in overcoming the major practical constraint of unavailability or poor persuasiveness of audit evidence, thus reducing auditor's dependency on client data (Yoon et al., 2015). For example, where sales forecasts are unconvincing or simply not made available to help auditors understand production volume and inventory levels, text analysis of news articles, product discussion forums and social networks may effectively inform about sales trends (Yoon et al., 2015). As a further example, "an auditor seeking to test the existence assertion relative to fixed assets would face a less-challenging task if each asset record were complemented with pertinent audio, video, and textual information" (Warren et al., 2015, p. 402). In a nutshell, Big Data analytics offers a great potential to assist the auditor in performing a broad spectrum of activities, thereby achieving high levels of efficiency and improving financial statements audits – which shall no longer need be based on samples – as Big Data analytics enables a comprehensive data analysis, with the additional benefit of reducing the number of *false positives* for further audit investigation (Cao et al., 2015, p. 427, emphasis added; Earley, 2015; EY, 2018; Richins et al., 2017). Such higher efficiency means reduced costs of audits and enhanced profitability (Alles, 2015). Considering the severe criticism toward the audit profession for their perceived failings ranging from the scandals of Enron (Alles, 2015; Enget et al., 2017), *Banco Privado Português*, *Banco Português de Negócios*, or *Banco Espírito Santo*, to the financial crisis, auditors may take advantage of Big Data as a way of increasing the effectiveness and credibility of their audits.

A paradigm shift in the auditing context is expected as a consequence of Big Data analytics, with sample-based auditing being replaced with "auditing by exception" since automated processes could direct auditors' attention to examine unusual patterns and anomalies (Richins et al., 2017, p. 63; Warren et al., 2015). For example, if analysis of social networks revealed a negative reputation of a certain product and its sales have increased, this inconsistency would be seen as a "red flag" (Yoon et al., 2015, p. 433). It has been argued that the most common use of data analytics in accounting is in assurance tasks, namely in fraud detection by identifying unusual data flows such as high-frequency or manipulated transactions, duplicate vendor payments, or abnormal transaction volumes (Schneider et al., 2015).

Activities in which Big Data analytics may assist the auditor include identifying and assessing risks at several levels (e.g., (i) accepting or continuing an audit engagement – bankruptcy or management fraud risks, among others; (ii) material misstatement in financial statements due to fraud; (iii) material misstatement through understanding the entity and its environment including evaluation of internal controls and testing their operating effectiveness), performing substantive analytical procedures in response to auditor's assessment of the risks of material misstatement, and performing analytical procedures near the end of the audit to assist the auditor in forming an overall conclusion about the financial statements (Cao et al., 2015; Earley, 2015; Yoon et al., 2015). Furthermore, by adding Big Data to the types of financial statement evidence that auditors typically analyze, these professionals can move from traditional audit work of *ex post* nature to the realm of predictive analytics, and, consequently, better assist their clients in strategic formulation, as well as predict errors or misstatements in accounts (Earley, 2015; Yoon et al., 2015). Arguably, increased efficiency in auditing activities will not decrease the need for auditors. By testing financial statements assertions on entire populations, the list of outliers requiring auditors critical thinking skills follow up is expected to increase exponentially; also as unstructured data is brought into analyses, new opportunities may emerge for auditors to provide assurance services over their veracity (Richins et al., 2017).

Yet, and despite some proactive partnerships between the Big Four accounting firms and technology-based firms (Fay & Negangard, 2017; Richins et al., 2017), current applications of Big Data analytics in external auditing (Alles, 2015; Cao et al., 2015; Earley, 2015) and accounting (Appelbaum et al., 2017) remain scarce. The reasons for the limited impact Big Data analytics have had on management decision-making, strategic analysis and forecasting include (i) shortage of appropriate hardware and software resources to handle computational challenges (Cao et al., 2015), which is especially true regarding small- and medium-sized enterprises (SMEs), (ii) implementation of ERP systems focus on improving the efficiency of financial reporting (Appelbaum et al., 2017), and (iii) lack of expertise on analytics and Big Data (Alles, 2015; Appelbaum et al., 2017; Cao et al., 2015). As for auditing, its highly regulated environment has also been identified as a discouraging factor toward the adoption of Big Data analytics (Earley, 2015). Though this is a matter of dispute with existing ISAs, which have been identified as the major facilitators of Big Data usage by auditors, to the extent that auditors are not currently exercising the full discretion that these standards give them (Alles, 2015). For example, according to ISA 200, audit evidence sources include nonfinancial data outside the client. Corroborating with and heading in the direction of Big Data, ISA 330 also has a very open view of what constitutes audit evidence (Alles, 2015).

Importantly, accounting and audit professionals should be mindful of breakdowns in information processing heuristics when facing excessive information from Big Data, which may compromise quality judgments (Enget et al., 2017). In this regard, McKinney et al. (2017) argue that accountants need to approach Big

Data analysis as informed skeptics, being ready to challenge the analysis by questioning the limits of measurement and representation, the subjectivity of insights, the challenges of statistics, and integrate datasets, as well as the effects of underdetermination and inductive reasoning. In the *era* of Big Data, analysts need not only the skills to analyze data, but also a higher-order thinking skill set, such as the ability to ask good questions and understand the limits of the analysis (McKinney et al., 2017).

4.4 THE CASE OF SMEs

It is well established that the potential benefits of Big Data extend to SMEs, favoring their competitive advantage and growth (e.g., Del Vecchio et al., 2018; Sen et al., 2016). Since most SMEs worldwide do not have their financial statements audited, accounting analytics may act as a proxy as it has the potential to offer the main benefits of external auditing: improved confidence in financial reporting and increased efficiency of business processes. Accounting analytics allows identifying trends and anomalies in businesses processes and controls, thereby creating opportunities for accountants to add value to their service provision to SMEs by recommending managers refinements to the internal control systems in place – the equivalent with the "Management letter" of external auditors. For example, in a purchase-to-pay process, accounting analytics tools may uncover price trending of items, calculate share of business by vendor or by item, and identify a wide range of possible anomalies – such as, duplicate purchase orders or invoices, purchase orders raised after invoice date, payments splits to avoid authority matrix, duplicate payments or payments in excess of invoice amount, time gap in posting invoices, among others (EY, 2018). Thus, by identifying problematic areas/activities where controls may have been overridden, it is possible to understand the reasons and establish required improvements (EY, 2018).

Despite the above, most SMEs suffer from a dearth of financial and organizational resources incompatible with handling the computational challenges of Big Data. ERP systems, able to cope with the Big Data Vs of volume and velocity, have become widespread in all but the smaller organizations (Alles, 2015). Therefore, SMEs are slow adopters of Big Data analytics (Coleman et al., 2016). Evidence suggests that Big Data success stories are mostly associated with larger companies, which benefit from more sophisticated human and organizational structures (Del Vecchio et al., 2018). In a more optimistic perspective, it is expected that SMEs will focus on the opportunities stemming from Big Data (Sen et al., 2016), namely because the adoption of Big Data technologies and approaches has been favored by cloud computing, whereby storage costs decreased exponentially (Del Vecchio et al., 2018). Some open sources and cost-effective solutions (e.g., Hadoop) have started to be configured as standards for storing, processing, and analyzing large and differentiated (un)structured datasets (Del Vecchio et al., 2018; Sen et al., 2016). Furthermore, SMEs are more flexible and prone to quick adaptation to changes toward efficiency than the largest (Sen et al., 2016). Yet, the lack of Big Data expertise has been identified as one of the six major challenges to SMEs growth (Sen et al., 2016), which reinforces the opportunities for accountants and auditors.

4.5 THE IMPACT ON ACCOUNTING EDUCATION

Accounting education objectives should reflect how accountants add organizational value (McKinney et al., 2017). Therefore, accounting (as broadly understood to include different specialties, namely auditing) curricula need to evolve and adapt to changes in the business world associated with moving into a data-driven decision-making paradigm. Ensuring that fresh graduates in accounting can work with Big Data is critical to adapt to the data-centric environment (Coyne, Coyne, & Walker, 2016; Dzuranin et al., 2018; Earley, 2015; Janvrin & Watson, 2017; Schneider et al., 2015; Sledgianowski et al., 2017; Yoon et al., 2015).

Given that data-driven evolution is transforming the accounting profession, recent initiatives have emphasized the importance of integrating Big Data and business analytics into the accounting curricula of Higher Education Institutions (HEIs). For example, AACSB (2016) emphasizes the interdisciplinary nature of data analytics, data management, and information technologies, and it recommends an accounting curriculum that exposes students to hands-on use of appropriate tools for Big Data and the complexities of decision-making (see also Janvrin & Watson, 2017). In particular, AACSB's Accreditation Standard A7 specifies that accounting degree programs should include learning experiences that develop skills and knowledge related to the integration of information technology in accounting and business. Included in these learning experiences is the development of skills and knowledge related to data creation, data sharing, data analytics, data mining, data reporting, and storage within and across organizations (AACSB, 2016, p. 30).

Consistently, PwC (2015, p. 4) points out that "today's accounting curriculum should be updated to equip students with new skills, especially in technology and data analytics". PwC (2015) also outlined recommendations to develop the technical skills in audit, tax, risk management, and consulting to use data analytics to obtain new insights and new solutions. In line with these concerns, in August 2014, at the annual meeting of the American Accounting Association, a panel session was held to discuss incorporation of more data analysis courses into accounting curricula (Earley, 2015).

In response to the relentless call for incorporating into the already loaded accounting curricula the technical foundations of data analytics and related skills, several academics have endeavored to suggest methods/approaches to facilitate such incorporation. A major challenge for accounting department administrators dwells on determining which data analytic skills and tools are relevant to the accounting profession, and where and how to include this subject in the curriculum (Dzuranin et al., 2018). For example, Sledgianowski et al. (2017) provide a method based upon Lawson et al.'s (2014) competency integration for the accounting education framework. The authors organize Big Data and business analytics instructional resources available (e.g., case studies and software tools) into the following three core groups of competencies laid down by Lawson et al. (2014): accounting, foundational, and broad management. McKinney et al. (2017) developed a framework and an illustrative example to facilitate accounting students to become informed skeptics in the *era* of Big Data by explaining the conceptual relevance of topical areas, such as the limits of measurement and representation, the subjectiveness of insight, the challenges of statistics and datasets integration, and the effects of underdetermination and inductive reasoning to Big Data analysis. In addition, example questions that accountants conducting Big Data analysis should be asking regarding each topic were identified. Furthermore, for each topic, references to additional resources were provided and students could have access to them to learn more about conducting an effective Big Data analysis. Based upon a broad exploratory survey of an accounting faculty, Dzuranin et al. (2018) found strong support for a hybrid approach whereby accounting study plans should include a stand-alone course emphasizing data analytic competencies alongside with accounting courses with data analytic competencies ingrained. Consistently with the established importance of critical thinking and judgment for the accounting profession (Enget et al., 2017; McKinney et al., 2017; Shah, Horne, & Capellá, 2012), Dzuranin et al. (2018) found accounting students perceived data-driven critical thinking skills, in other words, the ability to ask questions that can be answered using data, as the most important data analytics skills to be developed. Furthermore, the authors found hands-on projects and case studies to be the most desirable methods for delivering data analytics to students.

Both AACSB and PwC have been stressing the need for bachelor's degrees study plans to early expose accounting students to analytics, rather than courses on information technology and analytics being optional (AACSB, 2016; PwC, 2015; Rikhardsson & Yigitbasioglu, 2018). For example, the first year study plans leading to an accounting degree usually begin with a "financial principles" course, which may take advantage of technology, for instance, interactive data visualization, to engage students in the study of the more strategic aspects of accounting (Sledgianowski et al., 2017, p. 82). In the same line, PwC (2015, p. 14) contends that "Universities should infuse analytical exercises into existing curriculum to help students develop data analytics proficiency on top of their core accounting skills". Janvrin and Watson (2017) provide a review on resources to support instructors in promoting Big Data skills in their classrooms in order to better prepare accounting students for today's environment.

Big Data can play a central role in building bridges between classroom learning and the real world by bringing in real examples to students. Educators can use two key characteristics of Big Data – volume and variety – to quickly identify and select from a large dataset relevant real world disclosures to illustrate intermediate and advanced accounting principles (Sledgianowski et al., 2017). Accounting educators can enhance students' understanding by providing examples, demonstrations, and practical application, using software, tutorials, and case studies whenever feasible (Sledgianowski et al., 2017). For example, students in a cost accounting course could employ business analytics tools to develop an activity-based costing model and scorecard to communicate organizational goals and strategies. Similarly, an auditing course embeds plenty of opportunities for students to learn the impact of information systems and technology on the audit activity (Sledgianowski et al., 2017). These include researching auditing standards and techniques, conducting audits, analyzing datasets for fraud, and assessing internal controls. Fay and Negangard (2017) provide a case to utilize data analysis skills and harness the power of Big Data on financial statement audits – an analysis of journal entries for potential red flags of fraudulent financial reporting. Enget et al. (2017) provide a teaching case study to introduce accounting students to the concepts of Big Data and data analytics in conjunction with Section 316 of the American Institute of Certified Public Accountants (AICPA) Professional Standards (AU[1] Section 316, "Consideration of Fraud in a Financial Statement Audit"), and journal entry test work. In sum, there is a broad consensus that "Academics, as educators, certainly must revamp their accounting and auditing curricula to provide the necessary skills for Big Data in the accounting and auditing profession" (Griffin & Wright, 2015, p. 379).

4.6 CONCLUSION

The ubiquity of Big Data and business analytics for companies to remain competitive and attractive to external stakeholders in a globalized and technological world pose new challenges to the accounting and auditing professions that can be turned into opportunities. There is a broad consensus that the exponentially growing amount of data made available by developments in computing and telecommunications technology, leading to fundamental changes in the decision-making process, will be a disruptive force in accounting (Warren et al., 2015). Advances in data analytics are making it increasingly easier for non-accountants, such as computer scientists and other data scientists, to perform accounting and auditing tasks (Richins et al., 2017). Thus, if accountants and auditors do not quickly evolve into adopting Big Data analytics massively, existing technology-based firms with competitive advantages in data analysis (e.g. Google, Yahoo, Facebook, LinkedIn) could seize the opportunity to enter the already highly competitive accounting and auditing markets (Del Vecchio et al., 2018; Richins et al., 2017). The good news to these professionals is that businesses require assistance from those who besides being well versed in Big Data analytics also master business fundamentals – traits that being in short supply create unique opportunities for accountants and auditors to carve a niche for themselves by offering the rare combination of the two traits, rather than being left behind and replaced by data scientists (Earley, 2015; Richins et al., 2017).

We provided a review on reasons why accountants and auditors may eagerly embrace Big Data and data analytics. However, the slow pace of adjustment of auditors and accountants to the new realities of Big Data/business analytics, which most organizations today recognize as crucial to improve their businesses in many functions, is a matter of concern (Earley, 2015; Fay & Negangard, 2017; Griffin & Wright, 2015; PwC, 2015). To quickly address their gaps in expertise, both accountants and auditors may hire analytically trained professionals, or use the services of third-party solution providers for Big Data analytics, which raises privacy concerns (Cao et al., 2015), and squeezes profit margins in their services provision. So, this turns into a need for accountants and auditors to use Big Data, and then it may also be an unsurmountable obstacle for these professions to grab the related opportunities (Alles, 2015).

Current unfamiliarity of accountants and auditors with Big Data and related skills makes it difficult to *a priori* predict how effectively these professionals will embrace the emerging opportunities. Yet, sooner or later, one may arguably expect accountants and auditors to become Big Data analytically trained professionals as they have a history of expanding their skills and expertise (Janvrin & Watson, 2017; Richins et al., 2017). Moreover, they already master structured datasets easing the transition to working with unstructured data, and mastery of accounting grants a unique ability to comprehend and interpret the business environment (Richins et al., 2017).

Despite possibly taking time, the presence of Big Data will likely require more significant changes to the accounting and auditing professions than ever before, as we can expect a major overhaul of the accounting curricula to include more courses in programming, (un)structured databases, statistics, data visualization tools, among other Big Data analytics related skills (Alles, 2015; Earley, 2015; McKinney et al., 2017). Big Data has the potential to impact on long-term organizational decision-making and to fundamentally change the accountant's role in organizations, the use of accounting data, and the organization of the accounting function (Rikhardsson & Yigitbasioglu, 2018; Schneider et al., 2015). Furthermore, decisions in other functions shall also be affected as accounting data becomes broadly accessible through business intelligence and analytics solutions (Rikhardsson & Yigitbasioglu, 2018).

Despite several methods being suggested by academics to provide guidance to HEIs that are beginning the process of integration of Big Data and data analytics skills and tools into accounting curricula (e.g. Dzuranin et al., 2018; McKinney et al., 2017; Sledgianowski et al., 2017), and despite many organizations (e.g., EY's Academic Resource Center) have begun to provide HEIs with resources to be included in data analytics courses (Dzuranin et al., 2018), this remains a major issue. Current curricula of accounting and auditing bachelor's degrees offered by HEIs mostly overlook Big Data and analytics, as much as mandatory courses to gain entry to these professions. To encourage accountants and auditors to a comprehensive use of business analytics, professional bodies – chartered accountants and auditors' – could adjust their mandatory fields of knowledge accordingly. The already loaded curricula of accounting and auditing degrees to abide by the professions' entry requirements laid down by professional associations pose a major challenge to HEIs seeking to integrate into the study plans typical discipline-based accounting/auditing competencies with technological and information systems' competencies significant to Big Data and business analytics. Notwithstanding the increased sophistication of the data and the analytics, the most important aspect continues to be the human element, which includes critical thinking and judgment (Enget et al., 2017). In this vein, accounting education also has a role to play in training students to become informed skeptics – exercising the "Big Judgment" necessary to appropriately analyze Big Data (Earley, 2015; McKinney et al., 2017, p. 63; Shah et al., 2012).

Finally, SMEs are a vital part of economies worldwide (Sen et al., 2016). Thus, their financial reporting challenges have already been addressed by means of the simplified requirements of the IFRS for SMEs Standard when compared to full IFRS, considering the costs and the capabilities of SMEs to prepare financial information. To the contrary, international policy makers, information technology, business management, and data science communities have yet to promote SMEs uptake of Big Data analytics (Coleman et al., 2016).

NOTE

1. AU 316 represents a clarification of the Statement on Auditing Standards (SASs) No. 99 of the AICPA.

REFERENCES

Alles, M. G. (2015). Drivers of the use and facilitators and obstacles of the Evolution of big data by the Audit profession. Accounting Horizons, *29*(2), 439–449. https://doi.org/10.2308/acch-51067

Appelbaum, D., Kogan, A., Vasarhelyi, M., & Yan, Z. (2017). Impact of business analytics and enterprise systems on managerial accounting. International Journal of Accounting Information Systems, *25*, 29–44. https://doi.org/10.1016/j.accinf.2017.03.003

Association to Advance Collegiate Schools of Business (AACSB). (2016). Eligibility procedures and accreditation standards for accounting accreditation. Tampa, FL: Author.

Ballou, B., Heitger, D. L., & Stoel, D. (2018). Data-driven decision-making and its impact on accounting undergraduate curriculum. Journal of Accounting Education, *44*, 14–24.

Cao, M., Chychyla, R., & Stewart, T. (2015). Big data analytics in financial statement Audits. Accounting Horizons, *29*(2), 423–429. https://doi.org/10.2308/acch-51068

Chen, H., Chiang, R. H., & Storey, V. C. (2012). Business intelligence and analytics: From big data to big impact. MIS Quarterly, *36*(4), 1165–1188.

Coleman, S., Göb, R., Manco, G., Pievatolo, A., Tort-Martorell, X., & Reis, M. S. (2016). How can SMEs benefit from big data? Challenges and a path forward. Quality and Reliability Engineering International, *32*(6), 2151–2164. https://doi.org/10.1002/qre.2008

Coyne, J. G., Coyne, E. M., & Walker, K. B. (2016). A model to update accounting curricula for emerging technologies. Journal of Emerging Technologies in Accounting, *13*, 161–169.

Davenport, T. H. (2006). Competing on analytics. Harvard business review, *84*(1), 98–107.

Davidson, H. J., & Trueblood, R. M. (1961). Accounting for decision-making. The Accounting Review, *36*(4), 577–582.

Del Vecchio, P., Di Minin, A., Petruzzelli, A. M., Panniello, U., & Pirri, S. (2018). Big data for open innovation in SMEs and large corporations: Trends, opportunities, and challenges. Creativity and Innovation Management, *27*, 6–22. https://doi.org/10.1111/caim.12224

Dzuranin, A. C., Jones, J. R., & Olvera, R. M. (2018). Infusing data analytics into the accounting curriculum: A framework and insights from faculty. Journal of Accounting Education, *43*, 24–39. https://doi.org/10.1016/j.jaccedu.2018.03.004

Earley, C. E. (2015). Data analytics in auditing: Opportunities and challenges. Business Horizons, *58*, 493–500. https://dx.doi.org/10.1016/j.bushor.2015.05.002

Enget, K., Saucedo, G. D., & Wright, N. S. (2017). Mystery, Inc.: A big data case. Journal of Accounting Education, *38*, 9–22.

EY. (2018). Accounting analytics, https://www.ey.com/Publication/vwLUAssets/ey-accounting-analytics/$File/ey-accounting-analytics.pdf.

FASB. (1980). *Qualitative characteristics of accounting information*. Statement of Financial Accounting Concepts No. 2. Norwalk, CT: Author.

Fay, R., & Negangard, E. M. (2017). Manual journal entry testing: Data analytics and the risk of fraud. Journal of Accounting Education, *38*, 37–49.

Frey, C. B., & Osborne, M. A. (2013). The future of employment: How susceptible are jobs to computerisation? (pp. 1–79). Oxford, England: Oxford Martin School. Retrieved from https://www.oxfordmartin.ox.ac.uk/publications/view/1314

Granlund, M. (2011). Extending AIS research to management accounting and control issues: A research note. International Journal of Accounting Information Systems, *12*(1), 3–19.

Griffin, P. A., & Wright, A. M. (2015). Commentaries on big data's importance for accounting and auditing. Accounting Horizons, *29*(2), 377–379.

Maines, L. A., & Wahlen, J. M. (2006). The nature of accounting information reliability: Inferences from archival and experimental research. Accounting Horizons, *20*(4), 399–425.

Janvrin, D. J., & Watson, M. W. (2017). "Big Data": A new twist to accounting. Journal of Accounting Education, *38*, 3–8. https://doi.org/10.1016/j.jaccedu.2016.12.009

Lawson, R. A., Blocher, E. J., Brewer, P. C., Cokins, G., Sorensen, J., Stout, D. E., ... Wouters, M. J. F. (2014). Focusing accounting curricula on students' long-run careers: Recommendations for an integrated competency-based framework for accounting education. Issues in Accounting Education, *29*(2), 295–317. https://doi.org/10.2308/iace-50673

McKinney, E., Jr., Yoos, C. J., II, & Snead, K. (2017). The need for "skeptical" accountants in the era of big data. Journal of Accounting Education, *38*, 63–80.

Melville, A. (2015). International financial reporting – A pratical guide (5th ed.). Edinburgh Gate, UK: Pearson Education Limited.

PwC, P. (2015). Data driven: What students need to succeed in a rapidly changing business world. London: Author.

Richins, G., Stapleton, A., Stratopoulos, T. C., & Wong, C. (2017). Big data analytics: Opportunity or threat for the Accounting information? Journal of Information Systems, *3*, 63–79. https://doi.org/10.2308/isys-51805

Rikhardsson, P., & Yigitbasioglu, O. (2018). Business intelligence & analytics in management accounting research: Status and future focus. International Journal of Accounting Information Systems, *29*, 37–58.

Schneider, G. P., Dai, J., Janvrin, D. J., Ajayi, K., & Raschke, R. L. (2015). Infer, predict, and assure: Accounting opportunities in data analytics. Accounting Horizons, *29*(3), 719–742. https://doi.org/10.2308/acch-51140

Sen, D., Ozturk, M., & Vayvay, O. (2016). An overview of big data for growth in SMEs. Procedia – Social and Behavioral Sciences, *235*, 159–167. https://doi.org/10.1016/j.sbspro.2016.11.011

Shah, S., Horne, A., & Capellá, J. (2012). Good data won't guarantee good decisions. Harvard Business Review, *90*(4), 23–25.

Sledgianowski, D., Gomaa, M., & Tan, C. (2017). Toward integration of big data, technology and information systems competencies into the accounting curriculum. Journal of Accounting Education, *38*, 81–93.

Vasarhelyi, M. A., Kogan, A., & Tuttle, B. M. (2015). Big data in accounting: An overview. Accounting Horizons, *29*(2), 381–396.

Warren, J. D., Jr., Moffitt, K. C., & Byrnes, P. (2015). How big data will change accounting. Accounting Horizons, *29*(2), 397–407. https://aaajournals.org/doi/10.2308/acch-51069

Yoon, K., Hoogduin, L., & Zhang, L. (2015). Big data as complementary Audit evidence. Accounting Horizons, *29*(2), 431–438. https://doi.org/10.2308/acch-51076

Zhang, J., Yang, X., & Appelbaum, D. (2015). Toward effective big data analysis in continuous auditing. Accounting Horizons, *29*(2), 469–476.

5 Mobile Advertising Framework: Format, Location and Context

Bilal Aslam
School of Business and Economics,
University of Jyväskylä, Jyväskylä, Finland

Heikki Karjaluoto
School of Business and Economics,
University of Jyväskylä, Jyväskylä, Finland

CONTENTS

5.1 INTRODUCTION

Since the inception of smartphones in 2007, mobile advertising has advanced considerably. Mobile advertising has already surpassed all popular traditional media, such as television (TV), radio, and print advertising. Mobile advertising was estimated to become around $93 billion in 2019, surpassing even TV advertising spending by $20 billion (eMarketer, 2019). The popularity of mobile advertising is due to many factors, for example, functions like GPS (Global Positioning System), camera, and scanner are used more frequently in mobiles than in any other device, such as desktop computers (Mahmoud & Yu, 2006). Okazaki and Mendez (2013) attached features with mobile-like ubiquity. Mobile devices are continuously used or always turned on, and consumers can use them to search information. The Mobile Marketing Association (MMA) (2019) defines mobile marketing as a "set of practices" that includes "activities, institutions, processes, industry players, standards, advertising and media, direct response, promotions, relationship management, CRM, customer services, loyalty, social marketing, and all the many faces and facets of marketing." This definition considers mobile advertising a part of mobile marketing. Leppäniemi et al. (2004, p. 1) define mobile advertising separately as "any paid message communicated by mobile media with the intent to influence the attitudes, intentions and behavior of those addressed by the commercial messages." Meanwhile, according to Varnali and Toker (2010), there is no commonly accepted classification or framework for mobile marketing. This is partly because of the rapid advancement of wireless technology (Park et al., 2008) and continuously occurring changes in the industry. Leppäniemi et al. (2006) state that most studies on mobile marketing focus on user behavior and attitudes toward mobile marketing and mobile marketing effectiveness.

Industry-oriented advertising domains and other features that distinguish mobile from other devices and media remain unclear. Managers need to clearly understand the basic pillars and selling propositions of mobile advertising that cannot be attained by any other device or medium. Additionally, companies and advertisers cannot ignore privacy, which has become an important issue for all stakeholders (Okazaki et al., 2009).

The chapter will discuss the core factors responsible for making mobiles one of the most preferred advertising media. We will identify advertising domains that are deployable only in mobiles. These distinctive features and mobile-oriented advertising formats will help establish mobile advertising's authority over other devices and advertising media. In addition, this chapter will explore privacy, its influence on advertising effectiveness, and how it can be adapted such that laws are followed and no consumers are displeased. We will also incorporate the latest European Union (EU) privacy regulations (commonly known as General Data Protection Regulation [GDPR]) into the bigger spectrum of governmental and official steps to safeguarding consumer privacy.

For clarity, we will first identify and explain mobile-deployable advertising domains, including location-based advertising (LBA), in-application (in-app) advertising, SMS, and mobile social media and search engine advertising. Next, we will explain the concepts of location and context and their importance for mobile advertising. Regarding practical implications, we will discuss how companies can use location, context, and mobile-deployable domains in practice without breaching privacy regulations. Ultimately, this chapter aims to enlighten managers about mobile advertising and how it can be most effectively used.

5.2 RESEARCH METHOD

The scarcity of academic literature on this topic suggested that the search methodology for this review should proceed from the general to the specific—that is, from a broad to a specific classification of the niche topic of mobile and digital advertising. As is common practice for database searches, the keyword approach was used to identify previously published mobile advertising articles (Schibrowsky et al., 2007). The search began with the use of basic keywords, such as "mobile advertising" and

TABLE 5.1
Inclusion and Exclusion Criteria

Criterion	Inclusion	Exclusion
Study type	Peer-reviewed empirical and theoretical/conceptual studies; high-quality working/conference articles; and other Internet material, such as blog and video archives	
Language	English	Any other language
Sector	Private sector	
Date	2000 to 2018	Not directly relevant to mobile marketing or advertising Results directed toward marketing goals Smart TVs
Relevance	Any subdomain of digital advertising that enhances subject understanding, particularly of important related fields, such as mobile advertising, LBA, mobile social media, and privacy Industry-oriented information	Mobile Web advertising Mobile video advertising

"mobile marketing." After differences started to become apparent, we expanded our search by adding such keywords as "mobile social media," "mobile search engine advertising," "SMS advertising," "LBA," "location-based services," "privacy issues in mobile marketing," and "implication of GDPR."

Scientific literature (peer-reviewed journals), related Internet material (for recent development and statistics), and conference proceedings were searched for studies relevant to the topic. The search process was completed both horizontally (e.g., Google Scholar) and vertically (e.g., Science Direct, SAGE, Wiley, Springer, Emerald, JSTOR, IEEE, Taylor & Francis, Inderscience). The inclusion and exclusion criteria by Watson et al. (2017), as shown in Table 5.1, were adapted according to the requirement of this chapter.

On the basis of the inclusion and exclusion criteria (Table 5.1), 95 articles and conference proceedings were chosen of 150 articles found through the search. The time frame was 2000 onward, and only a few articles prior to 2000 were selected. Important conference proceedings included the Hawaii International Conference on System Sciences (HICSS). Most of the articles obtained via the vertical search were from the *Journal of Interactive Advertising*, *Telematics and Informatics*, *Marketing Science*, *International Journal of Mobile Communications*, *Electronic Commerce Research and Applications*, and *Journal of Research in Interactive Marketing*.

5.3 FINDINGS

Results showed that location and delivering messages according to consumer context are important features of mobile advertising. Grewal et al. (2016) explained that the uniqueness of mobile advertising lies in its ability to support location-based services

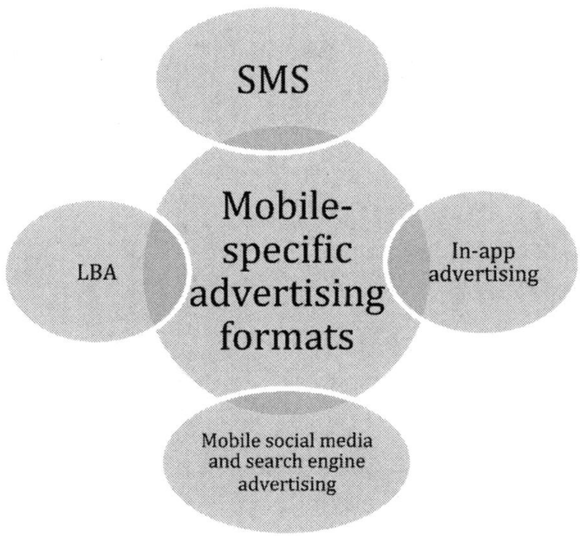

FIGURE 5.1 Classification of mobile-specific advertising media.

and in the important role of context in its effectiveness. Some advertising streams are meant for mobile advertising and mobile marketing only. These streams can be categorized into four main domains, namely, LBA, SMS, in-app advertising, and mobile social media and search engine advertising (Figure 5.1).

In the case of LBA, a desktop can be tracked and served with marketing and advertising communication. However, the role of desktop computers remains limited with regard to location-related advertising, as each is fixed in one location. By contrast, mobile devices are carried everywhere. Therefore, the full scale of location-based services is possible on mobiles only. In fact, nearly all previous studies discuss LBA with respect to mobile (Al Khasawneh & Shuhaiber, 2013; Hühn et al., 2017; Ketelaar et al., 2015; Unni & Harmon, 2007). With regard to LBA, we will attempt to understand the concepts of location and context in detail.

Figure 5.2 explains the mobile advertising framework. On the left side, the concept of location is discussed, and context in diverse practical situations. We then incorporated mobile-deployable advertising domains. LBA can be deployed through SMS, in-app advertising, and mobile social media and search engine advertising. Therefore, we attempted to present different possibilities of using LBA through these domains. Meanwhile, privacy is at the heart of mobile advertising, especially for mobile advertising methods that track consumers' location and send messages accordingly. Consumers may develop negative sentiments about an advertised brand if they think their privacy is threatened or violated. The emerging personalized and location-based mobile advertising, unless carefully monitored, may become an extremely intrusive practice (Cleff, 2007). Particularly after the implementation of GDPR in the EU, it has become absolutely important to understand privacy and governmental laws for smooth execution of mobile advertising. GDPR comprises laws meant to safeguard consumer privacy and prevent data misuse in the EU.

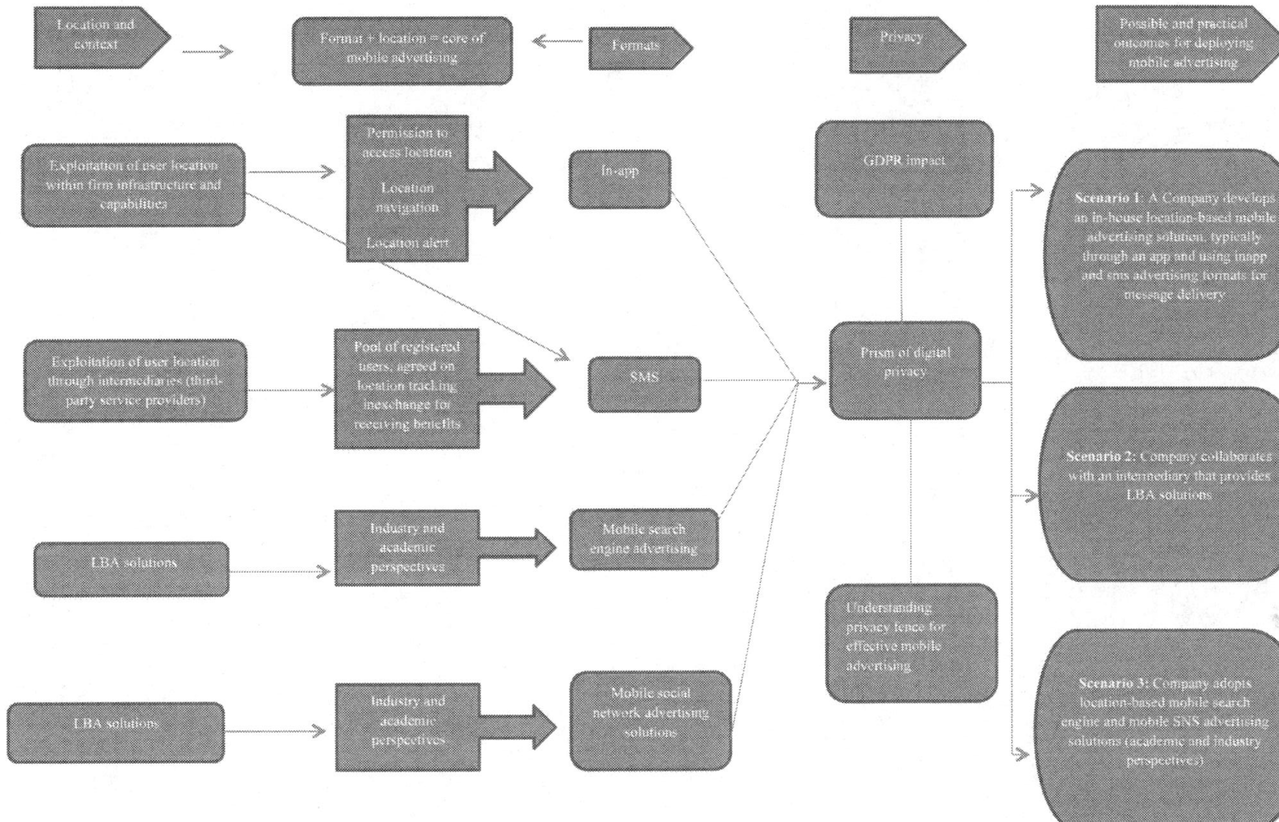

FIGURE 5.2 Mobile advertising framework, which explains the relationships between location, context, and mobile advertising formats. Then passing it through privacy regulations to come up with three practical cases that firms can consider to implement.

The intentions behind GDPR are simple—to keep abreast with the threat of cyber security in relation to strategy, legislation, and operations; to be ready to respond to such threats; and to ensure future resilience (Stormshield, 2017). We will discuss GDPR in a direction pertinent to our topics of interest only, that is, mobile advertising, privacy, and regulations regarding consumer or location data.

Every mobile marketing and advertising activity has to go through the prism of privacy. Thus, the prism of digital privacy, as depicted by Figure 5.2, mainly consists of consumers' basic privacy rights, such as being asked for consent for sending of advertising messages, and other governmental implications, especially GDPR.

5.3.1 Location-Based Advertising (LBA)

LBA is a location-based service. Such services primarily deal with determining user location. LBA is executed by attaching an advertising message to the determined location. Understanding the difference between context and location is very important. Context-aware LBA messages help convert prospects to actual sales. LBA messages might also create negative sentiments if they are not customized according to users' psychological and ongoing circumstances. With the adoption of contextualization data, users can receive convenient, compelling, and useful advertising (Simoes, 2009).

The basic merit of LBA lies in the fact that a user can be targeted according to his or her location (Unni & Harmon, 2007). Location congruency positively affects consumers' perceived relevance and the value of mobile advertisements (Hühn et al., 2017). Consumers exposed to LBA messages are likely to choose the advertised brands (Ketelaar et al., 2015). In the early days of LBA, it referred to marketer-controlled information customized for recipients' geographic positions and received via mobile communication devices (Bruner & Kumar, 2007). This definition helps in the understanding of LBA but does not include the concept of context, of which location is merely a small physical part. The other parts of context include time, people, community networks, and sensor-provided data (Fanjiang & Wang, 2017). In other words, location does not only correspond to a place in terms of its absolute location, it also needs to consider the position of other elements that are relevant to the service (Kjeldskov, 2007). Figure 5.3 shows that technical and consumer contexts are subparts of context. Location is at the core of all contexts, and in location, there are on-ground factors. The technical and consumer contexts were adopted from Grewal et al.'s (2016) mobile advertising effectiveness model. These factors, namely, technical, consumer context, and on-ground factors, should be considered before delivering an LBA message. When advertisers send mobile advertising messages to meet consumers' time, location, and preferences, consumer attitudes are positively affected (Al Khasawneh & Shuhaiber, 2013).

Technical and consumer contexts include any information that is available on the Internet, such as a consumer's activity on social media, upcoming important events, offline/online footprint, and places in the consumer's journey or search logs and history. For an example, a user's upcoming wedding anniversary is observed on social media. An LBA message sent prior to that event at the targeted location can increase the possibility of conversion to sales. In another example, online search logs show a consumer's interest in clothing. Sending an LBA message when the consumer is near

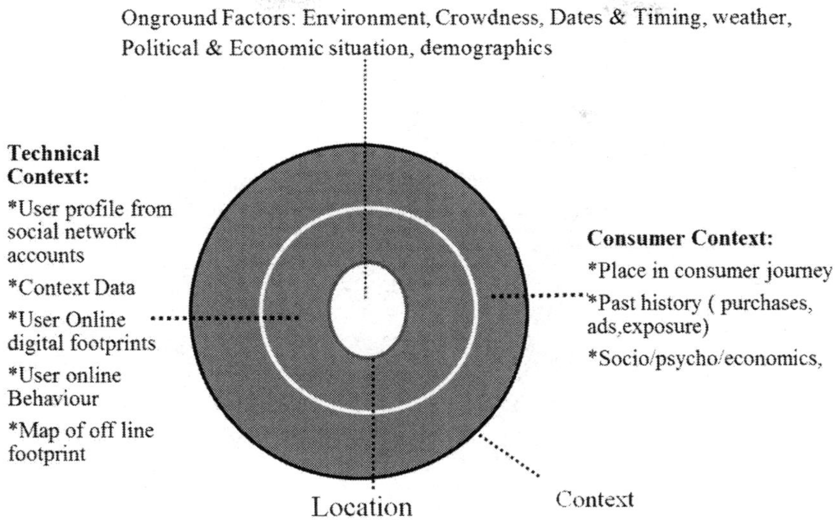

FIGURE 5.3 Relationship between location and different kinds of contexts and overall factors to be considered for improved LBA messages.

a physical store will be more beneficial for the clothing brand than selecting random people for its LBA campaign. The final example concerns a consumer who physically visits a retail shop, such as a grocery store, during particular days of the month. Sending an LBA message from the grocery retailer on those particular days offering new promotions and discounts will be beneficial. Thus, consumers' online activity, interests, or offline footprints can help in customizing LBA messages according to these individuals' needs.

On the other hand, ground factors are small elements that advertisers should consider at the time their LBA messages are intended to be delivered. For example, area conditions should be appropriate for shopping. Thus, LBA messages about shopping should not be sent when there are strikes or other unfavorable conditions at the store location. Consumers receive their salaries either during the beginning or the end of the month. Thus, selecting the appropriate date and time is also very important. All these factors can influence consumers' buying process. In addition, small on-ground factors at targeted locations must be considered before sending a location-based message; such factors include crowding or traffic that can hinder message processing (Wilson & Suh, 2017). However, overall targeting of consumers in crowded environments with mobile promotions may be beneficial (Andrews et al., 2016). Research shows that men are more receptive to LBA advertisements during work hours, and women respond well in their leisure time (Banerjee & Dholakia, 2012). Thus, it can be assumed that men should be targeted during office hours (typically 9 am to 5 pm) in all working days except public holidays and seasonal vacations. Meanwhile, women should be targeted after office hours, that is, prime time or during vacations and weekends. Marketers who are interested in time-specific campaigns, such as those involving discount coupons, should shorten coupon validity to signal urgency (Danaher et al., 2015).

Diverse factors can create positive or unfavorable attitudes toward LBA messages. For example, perceived utility, utilization of contextual information, perceived risks, and trust influence mobile users' perceived value of LBAs (Barnes, 2003; Lin et al., 2016). Richard and Meuli (2013) conducted experiments and concluded that LBA must be entertaining, informative, include some form of incentive, and not irritate consumers. Considerable customization, permission, and lack of perceived intrusiveness are keys to creating positive attitudes toward such messages and, ultimately, purchase intent (Gazley et al., 2015).

Privacy is at the heart of LBA. According to Banerjee and Dholakia (2012), LBA should be an opt-in process, given that sending commercial messages to wireless devices without recipients' permission is prohibited under the Telephone Consumer Protection in the United States (US) (Kimball, 2004). Furthermore, GDPR requires the sending of requests for consent in an intelligible and easily accessible form; an explanation about the purpose of data processing should be attached to this consent request.

5.3.2 SMS

SMS advertising refers to the transmission of advertising messages via mobile phones in the form of short text-based messages (Haghirian et al., 2005). Managers are interested in this medium because of the generally positive attitude of consumers toward SMS, its role in location-based services, and its effectiveness toward low-involvement products (Drossos et al., 2014; Okazaki & Taylor, 2008). SMS advertising is one of the most discussed topics in mobile advertising literature (Barwise & Strong, 2002)). The use of SMS is not restricted to its own domain. Other mobile marketing activities, especially LBA and even in-app advertising, can use SMS as a message delivery option. For example, SMS can be one choice for delivering an LBA message after the user location is determined. Bauer and Strauss (2016) confirmed this phenomenon, that is, SMS and MMS notifications are widely used for serving LBA messages.

Similar to LBA, SMS advertising is highly personalized and subject to privacy. Such media can attract negative sentiments and be perceived as highly intrusive. A study by Aydin and Karamehmet (2017) estimated that more than 60% of the respondents had negative attitudes toward SMS advertising. Hence, it is important to comprehend the factors that can help build positive consumer response toward SMS advertising. Properly addressing consumer privacy will also help reduce the perceived intrusiveness (Cortés & Vela, 2013). Furthermore, this basic privacy rule should be expanded. Consumers should be allowed to choose how many messages they are willing to receive in a given time frame (Cleff, 2007).

Like LBA, context plays an important role in the acceptance of SMS advertising. Context ranks as the most important driver of acceptance of SMS advertising (Dix et al., 2016). Barnes and Scornavacca (2004) showed that the irritation and interruption caused by SMS advertising can be reduced by advertisers by ensuring content relevance, providing incentives, and creating contextual congruency. Context-driven personalization technology strongly influences the persuasion process (Chutijirawong & Kanawattanachai, 2014). However, the way literature defines context in the case of SMS should be understood. In the early days of SMS, it was matching of the time, location, and message (Heinonen & Strandvik, 2003). By contrast, nowadays, consumer preferences and attitudes can be derived from social media or through online activities; physical footprints can be noted also. Therefore, the concept of context is no longer limited to time and location. The contextual model that we developed for LBA is also relevant to SMS advertising, especially in the case of online shopping or cases where consumers' online data (such as social media, search, and general browsing data) is available. Consumers' online activities on websites can also be easily monitored. Data regarding online activities may include a user's visited sections or products that are being considered. These elements can help interpret consumer psychology and needs. In particular, with the help of modern techniques, such as artificial intelligence (AI), managers can accurately predict consumer needs through data and send SMS messages accordingly.

In cases where such information is not available because of technological restrictions or other factors (e.g., a company has physical stores only or cannot obtain its consumers' online data), context can be properly addressed by creating consumer profiles. Such profiles should store small details, such as purchase history, buying trends, dates and times of purchases, and other demographic information. This information can be stored in real time as a consumer buys items from a brick-and-mortar store. This profile has to be updated whenever the consumer initiates any sort of exchange with the company, such as by making new purchases or asking for information. The SMS advertising should be aligned according to the details in the consumer's profile. Likewise, timing is valuable in contextualization. According to Luo et al. (2014), same-day SMS can be 9.5 times less effective as SMS that is sent one day before. The underlying reason could be the fact that consumers feel mentally prepared for the perceived benefits and potential costs of the offer. However, messages sent more than two days before the offer date are 71% less effective as one-day-prior SMS. The main challenge is deciding which advertisements to broadcast to which customers at what time, given all the information at hand (Reyck & Degraeve, 2003). Best effects occur on Mondays and weekends. Additionally, participants in a past work showed higher acceptance and purchase intention during the afternoon and evening than in the morning (Rau et al., 2011).

The third important factor that must be understood is the internal characteristics of the SMS messages themselves, particularly the alignment of textual information and the selection of words, that is, the message that should be written in text form to create positive consumer attitude. Barwise and Strong (2002) advised that SMS messages should be short, straight, funny and entertaining, and eye-catching; they should also offer some incentive, such as promotions. Muk and Chung (2015) concluded that perceived usefulness and social influence help create a positive attitude toward SMS advertisements. Moreover, small factors, such as interactivity, appeal, product involvement, and attitude toward SMS advertising in general, directly influence attitudes toward advertisements (Drossos et al., 2007).

5.3.3 In-App Advertising

Consumers throughout the globe have widely adopted mobile applications after the introduction of multi-touchscreen smartphones in 2007. Gupta (2013) explained that mobile applications fall into four categories, namely, games and entertainment, social networks, utilities (such as maps, clocks, calendars, cameras, and email), and branded apps. Generally, there are two ways of using mobile applications for marketing or advertising purposes. First, they can be used for in-app advertising across various types of mobile applications. Second, companies or brands can be in contact with their consumers through their own branded apps. By doing so, a company can do various activities, such as LBA, sales promotions, communication of important events, and tasks related to consumer engagement.

Most apps on Google Play and the Apple Store are free for download. The revenue model adopted by many free apps includes advertisements (advertisements) that are embedded in apps and displayed at various points during use (Rodriguez et al., 2012). The ecosystem of in-app advertising is explained through Figure 5.4. A developer creates an application, registers with an advertising provider, and uploads his or her work to an application platform (such as Google Play) for consumers to download. Consumers download the app for free, and in exchange for its usage, they are obliged to watch/interact with advertisements coming from an advertisement provider. To ensure that advertisements are delivered and displayed nonintrusively, advertisers usually have strict guidelines on how advertisements should be used and displayed in apps (Li et al., 2015). Other revenue models include offering paid added services. For example, in a gaming app, a user can buy items for the game to progress quickly in the game.

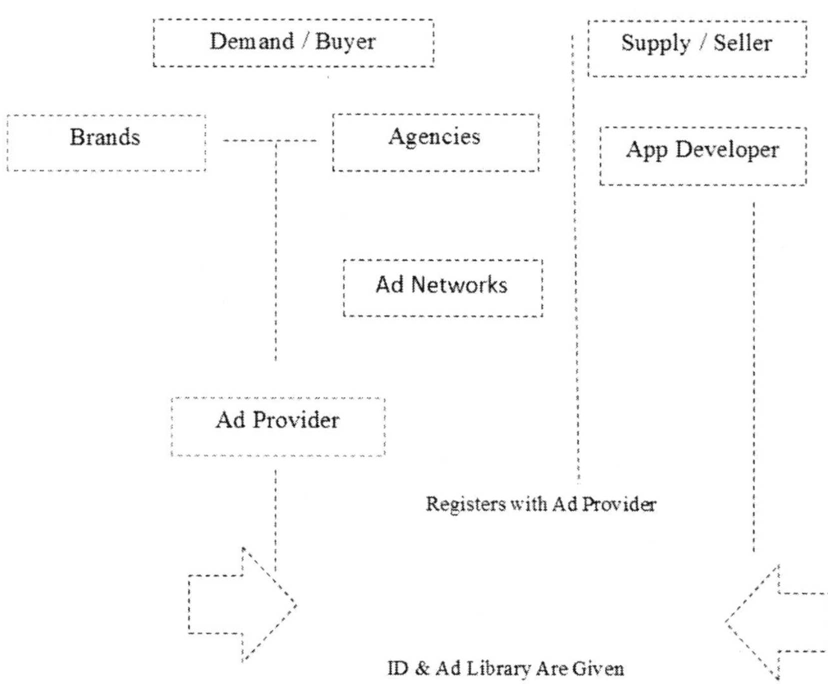

FIGURE 5.4 In-app advertising ecosystem and mechanics.

The attitude toward mobile app advertisements can be negative (Aydin & Karamehmet, 2017) primarily due to the fact that in-app advertising appears while a consumer is using the app. This setup can disturb customer usage, thereby eliciting a negative attitude toward the advertising. In the study of Ghose and Han (2014), app demand decreased with in-app advertisement that was shown to users while they were engaging with the app. Additional factors, such as distraction, interruption, delay, and stoppage, hinder in-app advertising (Nadeem et al., 2015). Nevertheless, some minor adjustable factors can increase in-app advertising effectiveness. Cicek et al. (2018) observed that users recall banners and their contents better when such banners have landscape orientation and are located at the top of apps. However, for a customer who uses an app for a long time or has been exposed to many advertisements before, the probability of clicking on advertisements will significantly diminish (Sun et al., 2017). It is also preferable to advertise in apps that satisfy different gratification factors, such as personal relationships and surveillance. These apps might include social media, gaming, and assistance apps (Logan, 2017).

Alternatively, mobile apps can be used for marketing and advertising purposes through branded apps, which are the own apps of companies or brands. This method is also highly effective for location-based services. Companies or brands do not need to work with third-party firms for advertising communication, unlike in the case of in-app advertising. Instead, these companies or brands develop their own apps and encourage consumers to download them. This technique offers a personalized touch to consumer communication. Consumers find branded apps interesting because they fulfill consumers' informational needs (such as product reviews, store addresses, and coupons) or entertainment needs (such as games and check-ins). Continuous use of branded apps advertisements increase spending?(Wang et al., 2016). By contrast, discontinued use of branded apps is associated with reduced future spending (Kim et al., 2015). Creative execution, perceived interactivity, and compatibility positively influence branded app effectiveness (Bellman et al., 2011; Kang et al., 2015). Global brands design their apps in a way that improves ease of use, understanding, and control and navigation through mobiles (Kim et al., 2013).

5.3.4 Mobile Social Media and Search Engine Advertising

Social media has changed the way people communicate and interact. Yadav et al. (2015, pp. 337–338) defined mobile social media as "all mobile marketing applications that enable the formation of user generated content." Consumers' interest in mobile social media lies primarily in social influence and feeling of belonging. Such social influence creates a bandwagon effect; a consumer will continue using mobile social media even if they are not interested in it (Zhou, 2014). Users' behavioral intention toward using mobile SNSs is strongly determined by social influence, mobility, habitual behavior, and critical mass (Nikou & Bouwman, 2014).

Marketing and advertising firms use mobile social media for branding, market sensing, relationship management, and content development (Bolat et al., 2016). Meanwhile, location-based apps are used by individuals to share information with their friends regarding their hangout places, such as cafes, restaurants, and other important tourist locations, in one specific area (Kaplan, 2012). Their friends see this information and can be interested to join them. Foursquare and Gowalla incentivize users to check in using game mechanics. On these platforms, users earn badges after accumulating a certain number of check-ins or gain special recognition if they have more check-ins at a particular location than all other users (Chang & Sun, 2011). Companies or brands have several means of attracting nearby foot traffic through such apps. Foursquare offers display advertising that targets users who are most likely to become customers based on the places that these individuals have visited (for instance, at similar places in another part of town) or on these users' search for something related to any particular business (for instance, "pizza") (Foursquare, 2019). Companies can also open up their brand pages and offer special discounts and promotions for users who check in at their stores.

Flow refers to the time spent on social media with users' full attention and involvement. Gao and Bai (2014) showed that among other factors, flow also determines the continuance intention of mobile SNS. For creating flow, mobile social media should provide quality information without any technical glitches and offer ways to expand users' social circles (Zhou, 2015; Zhou et al., 2010). Consumers also show addictive characteristics with mobile social media usage; such characteristics are not vividly present in other advertising formats. Frequent use of social media and large network sizes, which eventually result in increased social capital, are the strongest predictors of SNS addiction (Salehan & Negahban, 2013; Yang et al., 2016). Consumers who are addicted to mobile social media and show good flow tendencies might show heightened interest toward advertising messages. A company or a brand that is interested in launching a social media platform in the form of a branded game or planning to open a new brand page on mobile social media networks must create their technical and content design with flow and addiction factors in mind. Additionally, consumers' networks of relationships with people who live and work in a particular society can be referred to as these consumers' social capital. Users who have higher social capital show more interest in advertising and increased advertisement-sharing behavior with their networks (Li & Wang, 2014).

Other salient attributes of mobile SNS advertising include social, mobile convenience, and active control informativeness, entertainment, and irritation (Ha et al., 2014). In general, advertisers should work closely with mobile social network providers to identify high-engaging users. In particular, advertisers should pay attention to users who seek news and information from mobile social networks, use a considerable amount of mobile data, and/or frequently check in to publish their locations (Wu, 2016). Brands should establish close relationships with consumers and establish and increase attachment, commitment, closeness, cooperation, and understanding in their interactions with consumers (Lee, 2016).

5.3.4.1 Mobile Search Engine Advertising

In recent years, the use of search engines through mobile devices has increased substantially. Search engine companies display keyword-targeted advertising. On mobile phones, these advertisements are typically displayed on the top and bottom of pages and contain two to three sponsored results, with the bottom containing only one to three organic results (Murillo, 2017). Major search engines, such as Google, have also begun indexing and ranking websites based on these sites' mobile content instead of desktop experience, as was done historically (Selbach, 2017). This change will impact Google's organic search listings. In other words, obtaining a high ranking on Google now requires mobile-optimized versions of websites.

The advertisement mechanism of Google mobile search engine advertising is similar to that for desktop. Advertisers bid against targeted keywords. Google combines bids with multiple quality factors, such as expected click-through rates (CTRs), landing page and advertisement relevance, and expected impact of advertisement formats in calculating a score or rank for each advertisement (Aslam & Karjaluoto, 2017). However, given the small screens of mobile devices and the increased trend of conducting local searches on mobile, studies show that additional measures are needed to enhance the effectiveness of mobile search engine advertisements. Because of the small screens of mobile devices, users spend more time on the first three results on search engine query retrieval pages than on the following results (Kim et al., 2012). Hence, the top rankings are very important in mobile search engine advertising. Additionally, mobile search advertisements should have clear call to action. For example, a simple call to action, such as "Call us," can bring improved results for retail businesses that are implementing mobile search engine advertising (Goh et al., 2015). Ghose et al. (2012) also showed that ranking effects are high on mobile phones, suggesting high search costs; links that appear on the top of the screen are especially likely to be clicked on mobile phones. This increased cost can be justifiable by managers who are advertising local content, given that a link to a store located close to a user's home is highly likely to be clicked on a mobile phone. The text itself should be informative and entertaining and should not cause irritation (Murillo, 2017).

Due to the difficulty of entering text on mobile, query length is shorter in mobile devices than in desktops (Kamvar et al., 2009). Interestingly, two-term queries are more common than one-term queries on mobile devices despite that the former typically requires more effort from users. This is in line with the main hypothesis, that is, users prefer to be precise to avoid refing the query and being required to input added text (Yates et al., 2007).

5.4 PRIVACY AND APPLICATION OF GDPR

Users have an ultimate right over their information; if these rights are violated, they become reluctant to disclose personal information, will not respond to advertising offers, and may even seek stricter regulatory control over mobile advertising (Okazaki et al., 2009). At the government level, consumer rights, their privacy, and data safety are now priorities. The EU's new GDPR framework was developed to safeguard consumers from privacy and data leakage threats. Large fines are imposed on companies who violate such laws. Cate (2016) showed that people become increasingly likely to accept potentially invasive technology if they think its benefits will prevail over its potential risk. Moreover, consumers are likely to accept such offers when offered a monetary reward for their usage of data or location data (Gutierrez et al., 2018). Users should also be given an easy opt-out option from these services. Designing such systems with complex privacy settings, thus giving consumers increased control, will benefit all parties (Kelly et al., 2011).

Increased personalization of users' privacy settings can significantly enhance mobile marketing adoption and use (Gurău & Ranchhod, 2009). Finally, consumers should be able to decide about when, where, and how often they want to receive messages (Cleff, 2007).

5.5 THEORETICAL IMPLICATIONS

This chapter reviewed previous literature related to mobile advertising and marketing to bring clarity, given the fragmented nature of existing research on this topic. Because of the newness of the topic and the rapid changes that have occurred in mobile technology in the last decade, only a few studies have attempted to explain the broad practical picture of mobile advertising, which also satisfies industry needs. Consequently, the substantial impact that mobile advertising is generating in the real world is hardly reflected in previous studies. The present chapter is constructed in a way that covers all the important dimensions of mobile marketing and advertising. Each domain is analyzed in light of all valuable previous findings to produce a holistic and conclusive results, rather than abstract ones, against each discussed category.

5.6 PRACTICAL IMPLICATIONS

After examining all mobile-oriented formats and combining them with location-based services and advertising, we recommend three practical scenarios that take privacy into account. In Scenario 1, a company develops an in-house location-based mobile advertising solution typically through an app; this solution uses in-app and SMS advertising formats for message delivery. In this scenario, the company develops and manages its own branded app. The project can be outsourced to a third party, but managing this project internally can give the company enhanced control. However, execution through a third party will ensure smooth operations and reduced hassle for the focal company. Third-party firms usually have the appropriate expertise for the required operations and will be professionally and technically stronger than the focal company. Hebdry (1995) gave the following guide regarding outsourcing decisions. If contracting out is cheaper than doing the job yourself, then outsourcing is the better option. In this way, a company not only saves money through improved efficiency, but also gains effectiveness by focusing on their in-house expertise. Afterward, a company or brand will bring consumers on board by marketing its app from existing customers to potential customers and to users who will be interested to sign up for this app. The company needs to clearly state the benefits that users stand to gain by signing up and allowing the company to track their movements. The benefits should be engaging and attractive enough to compel consumers to sign up. For example, a consumer who downloads the company's app will receive on-the-spot notifications when he or she is near the location of purchase and will obtain special discounts. Consumers will then become informed about the company's offers, latest trends, and announcements regarding seasonal sales and new arrivals. Additionally, a company has to provide some additional monetary benefits that will only be available for the users of the app. Integrating attractive features into the app can also address this issue. Bonus points or virtual currency can be given after a few months of active app usage. Virtual currency can be made redeemable from the company's brick-and-mortar or online stores.

Consumers should be asked for permission during signing up to avoid breaching any privacy fence or any GDPR framework. Users should be informed that their location will be tracked to give them advanced services and monetary benefits. They should also be notified that they can opt out from such offers at any time. Additionally, customers should be given control over their privacy. Initially, they should be asked about small details, such as their interest in receiving on-the-spot location messages and their preferred time, date, location, or frequency of receiving such messages. Consumers should be informed that they can change these settings at any time while using the app. Increased monetary and other benefits should be offered to consumers who are willing to participate and interested to receive messages frequently. A company that is also interested in tracking customers' social media and other online activities for serving enhanced contextual messages must clearly mention this purpose in the beginning; users should also be given the option to unsubscribe from such offering at any time they wish.

In Scenario 2, a company collaborates with an intermediary that provides LBA solutions. In Scenario 1, a company or a brand can engage with their existing and potential customers by exploiting user location and delivering mobile advertising and marketing messages. However, this scenario vastly ignores the random foot traffic that is near the targeted location. Given privacy concerns by customers and governmental laws, it is not recommended to target random traffic in public places. Even if it were allowed, its future seems quite unclear. Every person walking near a market place could be blitzed with advertising messages from unknown brands or companies.

Scenario 2 will attempt to address this problem through an intermediary service. This location-based intermediary service incorporates both customers and brands simultaneously. Figure 5.5 explains these phenomena; the right side shows consumers from diverse backgrounds and demographics signing up with this intermediary. The registration process will be the same and aligns with the GDPR framework, as discussed in Scenario 1. Users will give their explicit permission for tracking their location and for receiving location-based messages; in return, they will receive some benefit, such as monetary rewards. They will also be given opt-out options and other privacy customization options, which they can change at any time during the service, thus having full control over their privacy. The intermediary will segment consumers internally, analyze their suitability, and deliver suitable LBA by brands or companies. Meanwhile, advertisers, brands, and companies will also register themselves with the intermediary and give their details about the demographical consumers they are targeting. For example, a 30-year-old female consumer is located near the shopping mall. The intermediary can look for an appropriate brand for this consumer (such as any female-oriented consumer brand) and then send an advertising message to her; the advertising content will have already been given to the intermediary by the brand.

This entire process can be automated on the basis of real-time bidding. That is, when the intermediary finds a suitable consumer at a location, it can ask for bids from different brands. The highest and most appropriate bidder will deliver their advertising content to the consumer at the location. This whole process will be completed in less than a second, as it happens in programmatic buying or real-time bidding.

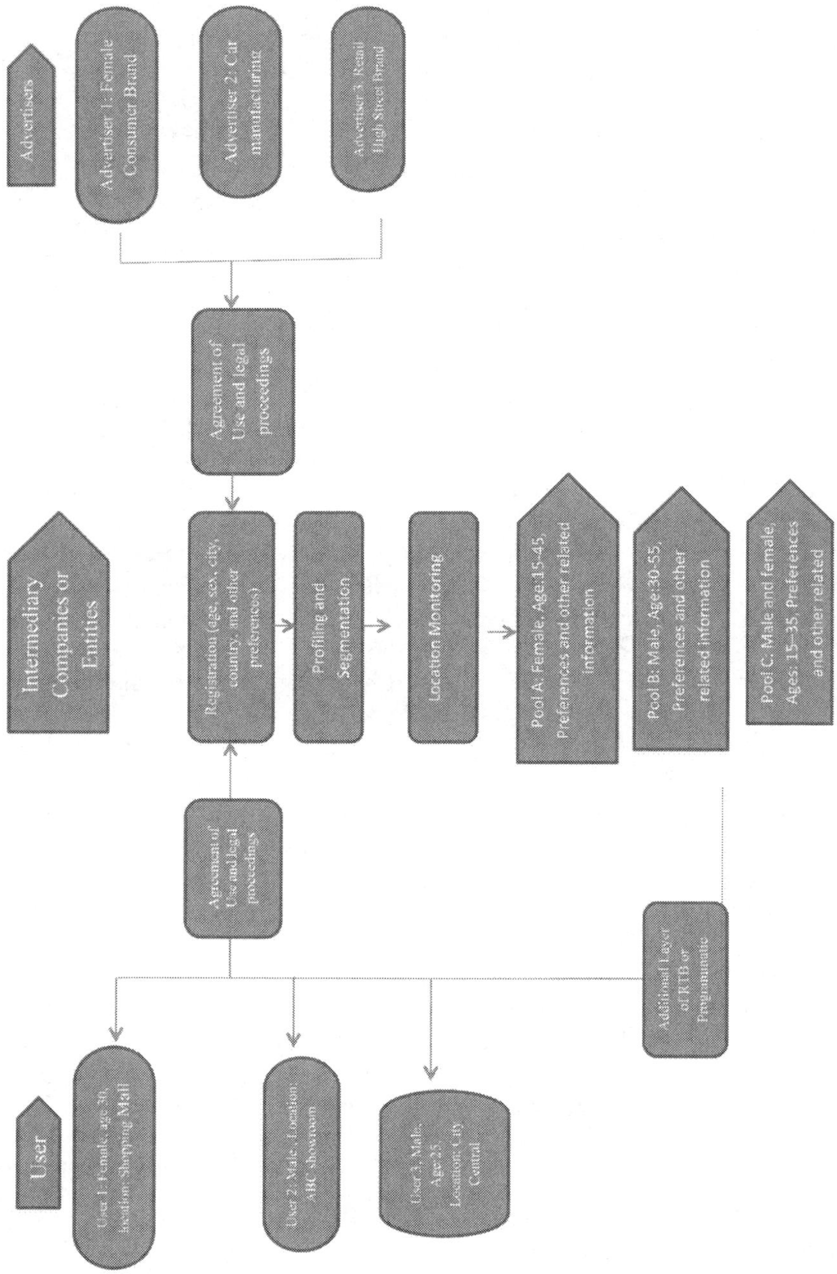

FIGURE 5.5 Hypothetical model explaining pool of consumers and advertisers or brands of an intermediary service.

Additionally, consumers can be contextualized by reviewing their online and social media activity and upcoming events.

In Scenario 3, a company adopts location-based mobile search engine and mobile SNS advertising solutions. Compared with Scenarios 1 and 2, Scenario 3 incorporates more industry-based advertising and marketing solutions. The term industry specifically means big Internet social media and search engine companies, such as Facebook, Foursquare, Instagram, and Snapchat. For simplicity, we can divide these companies in three broad categories. The first involves social media companies that are based on social interactions, such as Facebook and Instagram. The second includes social media companies that are based on location sharing, such as Foursquare. The third covers search engine companies, such as Google. All these services offer LBA and marketing solutions. By nature, when a consumer signs up for these services, they usually agree with all terms and conditions. Maintaining privacy and complying with governmental laws fall under in this domain. As an advertiser, manager, or a company will not be liable for any privacy intrusion or serious allegation, such as data breaches, advertising on these portals mainly happens in available spaces around different areas of their respective apps. These spaces are called Internet advertising paid slots and spaces (IAPS) (Aslam & Heikki, 2017). Depending upon the product and required results, an advertiser can choose a suitable option. For example, companies like Foursquare, Google Latitude, and Facebook Places are preferred for restaurants and brick-and-mortar stores, where consumers' online check-ins can help increase foot traffic or sales. Meanwhile, mobile search engine advertisements are good options for companies who wish to increase consumer contact through phone for further information or for booking services. Such companies believe that their potential consumer might search on their mobile first. Typical examples include car rental services. For general companies, such as those in Fast Moving Consumer Goods (FMCG) and similar areas, who are looking to build their brands and conduct general promotion of their physical stores, interaction-based social media apps are preferred, such as Facebook, Instagram, and Snapchat.

5.7 LIMITATIONS AND FUTURE RESEARCH DIRECTIONS

Although we attempted to discuss all advertising domains that can be deployed only in mobile, other advertising media, such as banner advertising on mobile-optimized advertisements and mobile video formats, were not covered primarily due to lack of academic literature about them. Thus, examining them as a separate domain was difficult. However, these domains are very important from the industry point of view. Mobile video advertisement spending in the United States will nearly double, going from almost $16 billion this year to almost $25 billion in 2022, and advertising on mobile video and on mobile-optimized websites will mostly be bought and sold pragmatically (eMarketer, 2019). Future studies on these topics will be valuable for the mobile advertising field. AI is also changing the future of digital advertising, and its application to mobile marketing is no exception. AI would be greatly beneficial for improved contextualization through big data analysis. This means that consumers will likely only see advertisements that matter to them. Therefore, future studies in this area will open up new avenues. Meanwhile, many studies have been conducted on social media advertising and search engine adverting with respect to desktop computers. However, are these findings pertinent to mobiles as well? In particular, mobile social media or search engine advertising done on the basis of location is a vastly unexplored field.

5.8 CONCLUSION

In this chapter, we highlighted the distinctive features of mobile advertising. Advertisers and marketers are becoming increasingly interested in spending on mobile advertising as it has already surpassed those for all other media, such as TV, print, and outdoor advertising in developed countries, especially the United States. We learned that because of the portability of mobile devices and the fact that users carry them everywhere, location is the prime reason among all important factors. Mobile advertising also gives a big opportunity to understand consumer context, which is hardly possible in any other device. We also identified certain advertising domains that are possible only on mobiles. By combining location, context, and these advertising domains, we attempted to understand mobile advertising's specialty and unique selling proposition. We then considered privacy and the implication of GDPR, as it affects mobile advertising, and recommended practical solutions for effective deployment of mobile advertising.

REFERENCES

Al Khasawneh, M., Shuhaiber, A. (2013). 'A comprehensive model of factors influencing consumer attitude towards and acceptance of SMS advertising: An empirical investigation in Jordan', International Journal of Sales and Marketing Management Research and Development, 3(2), pp. 1–22.

Andrews, M., Luo, X., Fang, Z., Ghose, A. (2016). 'Mobile ad effectiveness: Hyper-contextual targeting with crowdedness', Marketing Science, 35(2), 218–233, http://dx.doi.org/10.1287/mksc.2015.0905

Aslam, B., Karjaluoto, H. (2017). 'Digital advertising around paid spaces, E-advertising industry's revenue engine: A review and research agenda', Telematics and Informatics 34(8), 1650–1662.

Aydin, G., Karamehmet, B. (2017). 'A comparative study on attitudes towards SMS advertising and mobile application advertising', International Journal of Mobile Communications, 15(5), 514–536.

Banerjee, S., Dholakia, R. R. (2012c). 'Location-based mobile advertisements and gender targeting', Journal of Research in Interactive Marketing, 6, 198–214.

Barnes, S. J. (2003). 'Developments in the m-commerce value chain: Adding value with location-based services', Geography, 88(4), pp. 277–288.

Bauer, C., Strauss, C. (2016), 'Location-based advertising on mobile devices A literature review and analysis', Manag Rev Q (2016) 66:159–194 DOI 10.1007/s11301-015-0118-z

Barnes, S. J., Scornavacca, E. (2004), 'Mobile marketing: the role of permission and acceptance', International Journal of Mobile Communication, Volume 2 Issue 2, May 2004

Barwise, P., Strong, C. (2002). 'Permission-based mobile advertising', Journal of Interactive Marketing, 16(1), pp. 14–24.

Bellman, S., Potter, R. F., Hassard, S. T., Robinson, J. A., Varan, D. (2011). 'The effectiveness of branded mobile phone apps', Journal of Interactive Marketing, 25, 191–200.

Bolat, E., Kooli, K., Wright, L. T. (2016). 'Businesses and mobile social media capability', Journal of Business & Industrial Marketing, 31(8), 971–981.

Bruner, G. C., Kumar, A. (2007). 'Attitude toward location-based advertising', Journal of Interactive Advertising 7(2), pp. 3–15, https://doi.org/10.1080/15252019.2007.10722127

Cate, F. H. (2006). The failure of fair information practice principles. In Winn Jane K. (ed.). Consumer Protection in the Age of the 'Information Economy' (Markets and the Law). Surrey, UK: Ashgate Publishing, pp. 341–77.

Chang, J., Sun, E. (2011). 'Location3: How Users Share and Respond to Location-Based Data on Social Networking Sites', Proceedings of the Fifth International AAAI Conference on Weblogs and Social Media.

Chutijirawong, N., Kanawattanachai, P. (2014). 'The role and impact of context-driven personalization technology on customer acceptance of advertising via short message service (SMS)', International Journal of Mobile Communications, 12(6), https://doi.org/10.1504/IJMC.2014.064914

Çiçek, M., Eren-Erdoğmuş, İ., Daştan, İ. (2018). 'How to increase the awareness of in-app mobile banner ads: Exploring the roles of banner location, application type and orientation', International Journal of mobile Communications, 16(2), 153–166.

Cleff, E. B. (2007). 'Privacy issues in mobile advertising', International Review of LAW Computers & Technology, 21(3), 225–236.

Cortés, G. L., Vela, M. R. (2013). 'The antecedents of consumers' negative attitudes Toward SMS advertising: A theoretical framework and empirical, study', Journal of Interactive Advertising, 13(2), 109–117, https://doi.org/10.1080/15252019.2013.826553

Danaher, P. J., Smith, M. S., Ranasinghe, K and Danaher, T. S. (2015). 'Where, when, and how long: Factors that influence the redemption of mobile phone coupons, Journal of Marketing Research, LII, 710–725.

Dix, S., Jamieson, K., Shimul, A. S. (2016). 'SMS advertising the Hallyu way: Drivers, acceptance and intention to receive', Asia Pacific Journal of Marketing and Logistics, 28(2), 366–380, https://doi.org/10.1108/APJML-09-2015-0146

Drossos, D., Giaglis, G. M., Lekakos, G., Kokkinaki, F., Stavraki, M. G. (2007). 'Determinants of effective SMS advertising: An experimental study, Journal of Interactive Advertising, 7(2), 16–27, https://doi.org/10.1080/15252019.2007.10722128

Drossos, D. A., Kokkinaki, F., Giaglis, G. M., Fouskas, K. G. (2014). 'The effects of product involvement and impulse buying on purchase intentions in mobile text advertising', Electronic Commerce Research and Applications, 13 (2014), 423–430.

eMarketer. (2019). 'Mobile Trends 2019', https://www.emarketer.com/content/mobile-trends-2019

eMarketer. (2019). 'Mobile Video Advertising in 2019', https://www.emarketer.com/content/mobile-video-advertising-2019

Fanjiang, Y. Y., Wang, Y. Y. (2017). 'Context-aware services delivery framework for interactive mobile advertisement', Computer Communications, 99(1), 63–76.

Foursquare. (2019). 'About ads', https://support.foursquare.com/hc/en-us/articles/201066900-About-ads

Gao, L., Bai, X. (2014). 'An empirical study on continuance intention of mobile social networking services: Integrating the IS success model, network externalities and flow theory', Asia Pacific Journal of Marketing and Logistics, 26(2), 168–189.

Gazley, A., Hunt, A., McLare, L. (2015). 'The effects of location-based-services on consumer purchase intention at point of purchase', European Journal of Marketing, 49(9/10), 1686–1708.

Ghose, A., Han, S. P. (2014). 'Estimating demand for mobile applications in the new economy', Management Science, 60(6), 1470–1488.

Ghose, A., Ipeirotis, P., Li, B. (2012). 'Designing ranking systems for hotels on travel search engines by mining user-generated and crowd-sourced content'. Marketing Science 31(3), 493–520.

Goh, K. Y., Chu, J., Wu, J. (2015). 'Mobile advertising: An empirical study of temporal and spatial differences in search behavior and advertising response', Journal of Interactive Marketing, 30, 34–45.

Gupta, S. (2013). 'For mobile devices, think apps, not ads', Harvard Business Review,.

Gurău, C., Ranchhod, A. (2009). 'Consumer privacy issues in mobile commerce: A comparative study of British, French and Romanian consumers', Journal of Consumer Marketing, 26(7), 496–507, https://doi.org/10.1108/07363760911001556

Grewal, D., Bart, D., Span, M., Zubcsek, P. P. (2016), 'Mobile Advertising: A Framework and Research Agenda', Journal of Interactive Marketing, Volume 34, May 2016, Pages 3–14.

Haghirian, P., Madlberger, M., Tanuskova, A. (2005). 'Increasing advertising value of mobile marketing: An empirical study of antecedents', In Proceedings of the 38th Annual Hawaii International Conference on System Sciences, Washington, DC: IEEE Computer Society, https://doi.org/10.1109/HICSS.2005.311

Heinonen, K., Strandvik, T. (2003), 'Consumer responsiveness to mobile marketing', Stockholm Mobility Roundtable, academia.edu.

Hendry, J. (1995). 'Culture, community and networks: The hidden cost of outsourcing' European Management Journal, 13(2), 193–200.

Hühn, A. E., Khan,V. J., Ketelaar, P., Riet, J. V., Konig, R., Rozendaal, E., Batalas, N., Markopoulos, P. (2017). 'Does location congruence matter? A field study on the effects of location-based advertising on perceived ad intrusiveness, relevance & value', Computers in Human Behavior, 73, 659–668.

Kamvar, M., Kellar, M., Patel, R., Xu, Y. (2009). 'Computers and iPhones and Mobile Phones, oh my! A logs-based comparison of search users on different devices', WWW 2009 MADRID, April 20–24, 2009, Madrid, Spain.

Kang, J. Y. M., Mun, J. M., Johnson, K. K. P. (2015). 'In-store mobile usage: Downloading and usage intention toward mobile location-based retail apps', Computers in Human Behavior, 46 (2015), 210–217.

Kaplan, A. M. (2012). 'If you love something, let it go mobile: Mobile marketing and mobile social media 4x4', Business Horizons, 55(2), 129.

Ketelaar, P. E., Bernritter, S. F., Riet, J. V., Hühn, A. E., Woudenberg, T. J. V., Müller, B. C. N. & Janssen, L. (2015). 'Disentangling location-based advertising: The effects of location congruency and medium type on consumers' ad attention and brand choice', International Journal of Advertising, 36(2), 356–367.

Kim, E., Lin, J. Y., Sung, Y. (2013). 'To app or not to app: Engaging consumers via branded mobile apps', Journal of Interactive Advertising, 13(1), pp. 53–65.

Kim, J., Thomas, P., Sankaranarayana, R., Gedeon, T. (2012). 'Comparing scanning behaviour in web search on small and large screens', Proceedings of the Seventeenth Australasian Document Computing Symposium, ACM, New York, NY, pp. 25–30.

Kim, S. J., Wang, R. J. H., Malthouse, E. C. (2015). 'The effects of adopting and using a brand's mobile application on customers' subsequent purchase behavior', Journal of Interactive Marketing, 31, 28–41.

Kimball, R. (2004). 'FCC takes action to protect wireless subscribers from spam', News release of the Federal Communication Commission, http://hraunfoss.fcc.gov/edocs_public/attachmatch/DOC-250522A3.pdf

Kjeldskov, P. J. (2007). 'Understanding situated social interactions in public places: A case study of public, places in the city', Computer Supported Cooperative Work, 17(2–3), 275–290, https://doi.org/10.1007/s10606-007-9072-1

Lee, Y. C. (2016). 'Determinants of effective SoLoMo advertising from the perspective of social capital', Aslib Journal of Information Management, 68(3), 326–346, https://doi.org/10.1108/AJIM-10-2015-0155

Leppäniemi, M., Karjaluoto, H., Salo, J. (2004). 'The success factors of mobile advertising value chain', The E-Business Review, 4(1), 93.

Leppäniemi, M., Sinisalo, J., Karjaluoto, H. (2006). 'A review of mobile marketing research', International Journal of Mobile Marketing, 1(1), https://www.researchgate.net/publication/278023200_A_Review_of_Mobile_Marketing_Research

Li, W., Li, H., Chen, H., Xia, Y. (2015). 'AdAttester: Secure online mobile advertisement attestation using trust zone', MobiSys'15, May 18–22, 2015, Florence, Italy, ACM, http://dx.doi.org/10.1145/2742647.2742676

Li, Y., Wang, K. (2014). 'What affects the advertising sharing behavior among mobile SNS users? The relationships between social capital, outcomes expectations and prevention pride', PACIS 2014 Proceedings, 77, http://aisel.aisnet.org/pacis2014/77

Lin, T. T., Paragas, F., Bautista, J. R. (2016). 'Determinants of mobile consumers' perceived value of location-based advertising and user responses', International Journal of Mobile Communications, 14(2),

Logan, K. (2017). 'Attitudes towards in-app advertising: A uses and gratifications perspective', International Journal of Mobile Communications, 15(1), 2–22.

Luo, X., Andrews, M., Fang, Z., Phang, P. C. (2014). 'Mobile Targeting', Management Science, 60(7), 1738–1756, https://doi.org/10.1287/mnsc.2013.1836

Mahmoud, Q. H., Yu, L., 2006. 'Havana agents for comparison shopping and location-aware advertising in wireless mobile environments', Electronic Commerce Research and Applications, 5, 220–228.

Mobile Marketing Association (MMA). (2009). 'MMA updates definition of mobile marketing', https://www.mmaglobal.com/news/mma-updates-definition-mobile-marketing

Muk, A., Chung, C. (2015). 'Applying the technology acceptance model in a two-country study of SMS advertising', Journal of Business Research, 68(1), 1–6.

Murillo, E. (2017). 'Attitudes toward mobile search ads: A study among Mexican millennials', Journal of Research in Interactive Marketing, 11(1), 91–108.

Nikou, S., Bouwman, H. (2014). 'Ubiquitous use of mobile social network services', Telematics and Informatics, 31(1), 422–433.

Okazaki, S., Li, H., Hirose, M. (2009). 'Consumer privacy concerns and preference for degree of regulatory control', Journal of Advertising, 38(4), 63–77, https://doi.org/10.2753/JOA0091-3367380405

Okazaki, S., Mendez, F. (2013). 'Perceived ubiquity in mobile services', Journal of Interactive Marketing, 27(2), 98–111.

Okazaki, S., Taylor, C. R. (2008). 'What is SMS advertising and why do multinationals adopt it? Answers from an empirical study in European markets', Journal of Business Research, 61(1), 4–12.

Park, T., Shenoy, R., Salvendy, G. (2008). 'Effective advertising on mobile phones: A literature review and presentation of results from 53 case studies', Behaviour & Information Technology, 27(5), 355–373, http://dx.doi.org/10.1080/01449290600958888

Rau, P. L. P., Zhang, T., Shang, X., Zhou, J. (2011). 'Content relevance and delivery time of SMS advertising', International Journal of Mobile Communications, 9(1), 19–38.

Reyck, B. D., Degraeve, Z. (2003). 'Broadcast scheduling for mobile advertising', Operations Research, 51(4), 509–517, https://doi.org/10.1287/opre.51.4.509.16104

Richard, J. E., Meuli, P. G. (2013). 'Exploring and modelling digital natives' intention to use permission-based location-aware mobile advertising', Journal of Marketing Management, 29(5–6), 698–719.

Rodriguez, N. V., Shah, J., Finamore, A., Grunenberger, Y., Haddadi, H., Papagiannaki, K., Crowcroft, J. (2012). 'Breaking for commercials: Characterizing mobile advertising', IMC'12, November 14–16, Boston, MA.

Salehan, M., Negahban, A. (2013). 'Social networking on smartphones: When mobile phones become addictive', Computers in Human Behavior, 29, 2632–2639.

Schibrowsky, J. A., Peltier, J. W., Nill, A. (2007). 'The state of internet marketing research: A review of the literature and future research directions, European Journal of Marketing, 41(7), 722–733.

Selbach, J. (2017). 'All you need to know about Google's web crawler Googlebot', https://blog.seoprofiler.com/googles-web-crawler-googlebot/

Simoes, J. (2009). 'A generic enabler framework for contextualized advertising within interactive multimedia services', In IEEE International Symposium on a World of Wireless, Mobile and Multimedia Networks & Workshops, WoWMoM, June 15–19, pp. 1–3.

Stormshield, J. K. (2017). 'With GDPR, preparation is everything', Computer Fraud & Security, Computer Fraud & Security, 2017(6), 5–8.

Sun, Z., Dawande, M., Janakiraman, G., Mookerjee, V. (2017). 'Not just a fad: Optimal sequencing in mobile In-app advertising', Information Systems Research, 28(3), 511–528.

Unni, R., Harmon, R. 2007. 'Perceived effectiveness of push vs. pull mobile location-based advertising', Journal of Interactive Advertising, 7(2), 48–71.

Varnali, K., Toker, A. (2010). 'Mobile marketing research: The-state-of-the-art', International Journal of Information Management, 30(2), 144–151.

Wang, B., Kim, S., Malthouse, E. C. (2016). 'Branded apps and mobile platforms as new tools for advertising', In R. Brown, V. Jones, b. m. Wang (Eds.), The New Advertising: Branding, Content, and Consumer Relationships in the Data-Driven Social Media Era (pp. 123–156), Santa Barbara, CA: ABC-CLIO.

Watson, R., Wilson, H. N., Smart, P., Macdonald, E. K. (2017). 'Harnessing difference: A capability-based framework for stakeholder engagement in environmental innovation', Journal of Product Innovation Management, https://doi.org/10.1111/jpim.12394

Wilson, R., Suh, T. (2017). 'Advertising to the masses: The effects of crowding on the attention to place-based advertising', International Journal of Advertising, https://doi.org/10.1080/02650487.2017.1331967

Wu, L. (2016). 'Understanding the impact of media engagement on the perceived value and acceptance of advertising within mobile social networks', Journal of Interactive Advertising, 16(1), 59–73.

Yadav, M., Joshib, Y., Rahmanc, Z. (2015). 'Mobile social media: The new hybrid element of digital marketing communications', Procedia—Social and Behavioral Sciences, 189, 335–343.

Yang, S., Liu, Y., Wei, J. (2016). 'Social capital on mobile SNS addiction: A perspective from online and offline channel integrations', Internet Research, 26(4), 982–1000, https://doi.org/10.1108/IntR-01-2015-0010

Yates, R. B., Dupret, G., Velasco, J. (2007). 'A Study of Mobile Search Queries in Japan', In Proceedings of the Sixteenth International World Wide Web Conference (WWW2007), May 8–12, 2007, Banff, Alberta, Canada.

Zhou, T. (2015). 'The effect of network externality on mobile social network site continuance', Program, 49(3), 289–304.

Zhou, T., Li, H. (2014). 'Understanding mobile SNS continuance usage in China from the perspectives of social influence and privacy concern', Computers in Human Behavior, 37, 283–289.

Zhou, T., Li, H., Liu, Y. (2010). 'The effect of flow experience on mobile SNS users' loyalty', Industrial Management & Data Systems, 110(6), 930–946, https://doi.org/10.1108/02635571011055126

6 Marketing Analytics: Why Measuring Web and Social Media Matters

Xabier Martínez-Rolán
Faculty of Social Science and Communication,
University of Vigo, Pontevedra, Spain

Teresa Piñeiro-Otero
University of Coruña, Coruña, Spain

CONTENTS

6.1 INTRODUCTION: WHAT YOU CAN'T MEASURE, DOESN'T EXIST

According to the New Physics, what we can't measure cannot physically exist. This statement, attributed to American philosopher William Pepperrel Montague (Holt et al., 1912), is supposed to be the origin of the current motto "what cannot be measured, cannot be managed" (Morell, 2015).

From this realist trend, British physicist and mathematician Lord Kelvin also pointed out that what cannot be defined, cannot be measured, because "when you cannot measure it, when you cannot express it in numbers, your knowledge is of a meagre and unsatisfactory kind" (Kelvin, 1883, p. 17).

Indicators, KPIs (Key Performance Indicators), indexes, ratios, metrics—these are all elements that help measure performance in the analogue marketing world.

Of course, the world of digital marketing is familiar with these terms. In fact, in digital media, everything is likely to be measured, and with accuracy that is sometimes envied by conventional media (the press, radio, and television). While measuring audiences in tradition media is based on estimates against a sample of users, audiences in digital media are measured using the data of the total census of users.

According to the WAA Standards Committee (2008), "Web analytics is the measurement, collection, analysis and reporting of web data for purposes of understanding and optimizing web usage". At a time when any activity is measurable, with lots of data produced by digital analytics tools, the two fundamental questions are what to measure and how.

According to Google (n.d.), whenever we have to deploy a web analytics methodology—in trying to decide what to measure—we need to establish business objectives, conversion, and micro-conversion rates.

6.2 SETTING OBJECTIVES AND KPIs: THE SMART RULE

Setting objectives is one of the key aspects to consider when trying to measure the effectiveness of online marketing actions. While they have to be defined and prioritized by the website or social media owner, the web analytics manager can help the owner define the objectives.

As a rule of thumb, objectives need to be set according to the following three criteria:

- Visibility objectives: these are objectives linked to communication and they have an impact on the brand image. They try to achieve higher brand impact in terms of knowledge, brand recognition, creation or increase of the fan base, etc. These are important objectives but are sometimes difficult to measure in quantitative terms.
- Sales objectives: they try to get the user to do something on the web, it does not have to be direct sales, it can be downloading an app, that a user starts using your service, etc.

  These objectives are not only for the private sector, as some may erroneously think. The public administration, for example, can also set sales objectives: although they do not "sell" products directly, they do offer services. For example, increasing the users of a service against the preceding year can be a sales objective, as well as covering the number of seats in a training funded by the Administration.
- Loyalty objectives: they try to get the user to repeat a purchase. Another loyalty objective has to do with the user of a product becoming a voluntary brand prescriber, so that they become, literally, a brand fan.

In this sense, in order to follow-up and measure the quality outcomes of a strategy, we need to set the appropriate goals. Objectives should be set according to the

SMART rule proposed by Doran (1981) and totally relevant today in a fully quantifiable space such as the digital world.

According to Doran, each goal should be characterized by the following five attributes:

- Specific: Objectives have to clearly define what you aim to achieve.
- Measurable: Objectives have to have indicators to clearly measure efficiency.
- Attainable: Objectives have to be reasonable in terms of achievement.
- Relevant: Objectives have to be interesting.
- Time-Related: Objectives need to be set for a specific time frame.

On a webpage, there are multiple objectives one can try to achieve, such as a particular number of visits, retrieving a specific number of contact forms, downloading of an app or material embedded on the web (a handbook, a catalogue, etc.), a particular sales figure, or a certain number of subscriptions to a blog or newsletter.

On social media, setting objectives is more complex: they can involve diverting traffic to a website, answering questions about a product or service, creating a brand image, fostering customer loyalty through content marketing, the bidirectional character of social media requires paying more attention to objectives linked to product communication.

The link between web analytics and objectives are KPIs that allow us to see whether the objectives are being accomplished (Figure 6.1). The WAA Standards Committee (2008) indicates that "a KPI is infused with business strategy—hence the term, 'Key'—and therefore the set of appropriate KPIs typically differs between site and process types" (p. 3).

Each KPI is a metric, but not all metrics are KPIs. KPIs are the most important metrics, those directly linked with the objectives set. While the KPIs are strategic, the metrics are "dummy" and, often, vain.

In fact, the term "vanity metrics" is used for those that are easy to measure or monitor but might convey a false impression of growth (for example, on web environments, visit volumes; or in the social environment, number of fans on a Facebook fan page). These metrics tend to return an image that is not real; therefore, they have to be properly understood. They are like a distorted mirror in which the web or social media profile owners can see themselves magnified by having a large and/or active

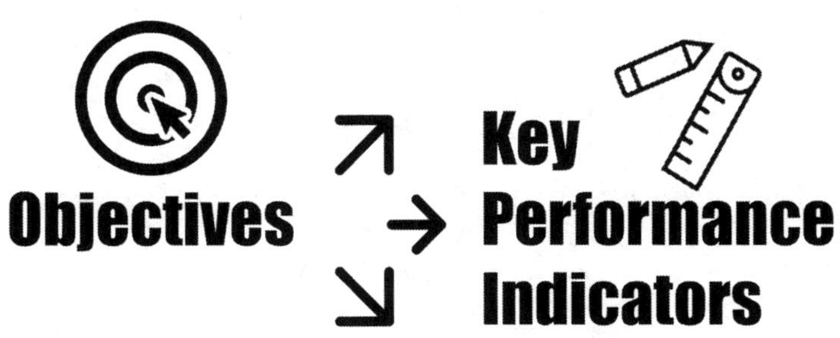

FIGURE 6.1 Objectives and KPIs. (Source: Prepared by the authors.)

community, even when this does not affect the sales volume, conversion rates, loyalty, or any of the objectives set.

KPIs vary depending on the type of web or social media platform and, above all, they also vary depending on the objective pursued. Let us look at some examples for different objectives:

Sales KPIs

- Abandoned cart rates
- Average order size
- Average order margin

Customer service KPIs

- Number of incidences received versus other matters or interactions
- Average time to close an incidence
- Assessment of response through the customer service web or social media

Social media KPIs

- Visibility: traffic figures
- Interaction or engagement: of users in any form
- Influence: analyzing our followers and their scope
- Popularity: subscriptions, followers, etc.

According to Kaushik (2011), what really matters is everything that happens after the post, tweet, or participation, and he therefore sets four KPIs that are widely used in social media analytics.

1. Conversation rate: this is the percentage of audience comments (or replies) per post. This is an important metric because the conversation is one of the interactions that require more effort (more than applause or amplification).
2. Amplification rate: the rate at which your followers take your content and share it through their network (number of retweets per tweet, number of shares per post on Facebook, etc.)
3. Applause rate: the rate at which your followers publicly like your content (number of loves on Instagram, number of likes on Facebook, etc.)
4. Economic value: this metric is assessed through macro and micro conversions. For that, Avinash suggests identifying social traffic to a website and its direct relationship to micro conversions with the help of web analytics. For example, the metric "per visit goal value" (economic value delivered by visitors from social media channels across my macro and micro conversions) can be useful to find economic value.

Each web and social media platform can set its own KPI, always in line with the previously set goals. They are very important in the development of a digital strategy; they should therefore be identified and detailed in a precise way.

6.3 FUNNEL ANALYTICS: CONVERSION FUNNEL

Any web-based activity leaves a trace: the browser, operative system and device you use, the incoming web data, the time you spend on the web, and content you consume – the consumed digital footprint of web content is deep, although not all data is interesting for the business owner.

Let us now look at e-commerce as an example. The business owner needs to generate sales; therefore, not all measurable information through web analytics is interesting.

E-commerce—just like any other website—needs conversions, that is, a visitor completing a target action. An e-commerce conversion can be a sale. Conversions are understood in terms of goals set: they can be clicking on an advertisement, registering for more information, starting a check out process, etc.

In this context, the term "funnel conversion" becomes key for digital marketing. It defines the different steps a user takes since they access a web service, an app, a social media site until they fulfil the objective set for that platform. It is the service owner who determines what the objective is: registering on a web, a purchase, a subscription to a newsletter, downloading an app, etc.

In essence, this simply allows the organization to determine the different steps that they must give users to achieve a web conversion.

For example, the funnel for ecommerce could entail four steps: accessing the website, visiting the product's page, adding the product to the basket, and completing the checkout process (Figure 6.2).

Many users access a website but not many visit a product page, and still fewer add the product to their basket, while only a limited few complete the order. This sequence has a funnel shape, and therefore the name.

The role of the conversion funnel is key to determine up to what extent there are losses in each step of the way until the final goal is accomplished. This is not a progressive tunnel, but in some of the phases we find critical points that need to be optimized.

Through this technique, one can detect potentially weak points and act on them.

For example, if the percentage of "add to cart" sessions differs much from the "transaction" sessions, we should ask ourselves what happens with the basket so that

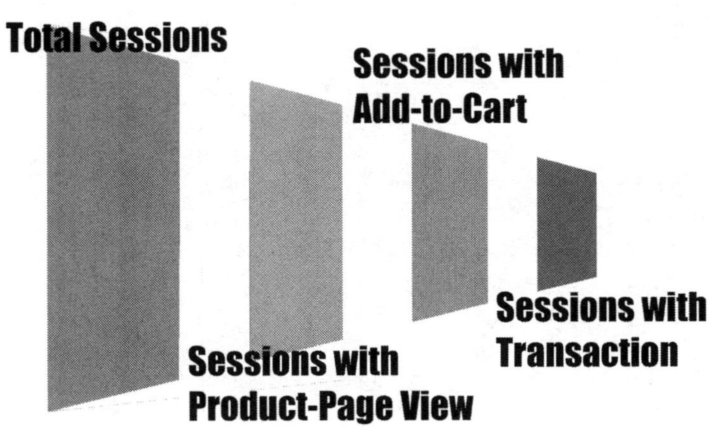

FIGURE 6.2 Conversion funnel in ecommerce. (Source: Prepared by the authors.)

the purchase is not completed: Is the process clear? Is this a simple process? Are the mailing and other costs clearly explained? Without the conversion funnel data, the website cannot be optimized.

6.4 MEASURING

Constant measuring means identifying and understanding the metrics, a step needed to compile the data produced by web analytics and social media. As KPIs are based on metrics, we need to know the most useful metrics to measure web and social media performance.

6.4.1 Web: Main Metrics with Web Analytics: Segments, Filters

As any web visit produces many data, we need to collect some basic metrics through web analytics, some about the user, some about the visit (or session), the pages visited, the time used, and the content consumed.

The data are obtained through a third-party DoubleClick cookie, Android Advertising ID and/or iOS Identifier for Advertisers (IDFA) (Google, 2019). In this way, the technical data related to the session are added to the demographic and interest information available in the cookies and information associated with the users' app activity.

These are the basic metrics per activity for web analytics:

- User: a user, for Google Analytics, is a combination of a unique random number and the first-time stamp (Sharma, 2016). A user can be new or recurrent. A new user is the first time a cookie registers them, while a recurring user is the person who comes a second or more times after being registered by the cookie.
- Visit / session: this refers to any activity undertaken by a user from the moment they access the web until they leave. There are two cases in which the visit might not be reliable: when the user spends more than 30 minutes inactive on the site, or even while browsing within those 30 minutes, the clock gets to 00:00, as the cookie runs out at midnight.
- Pages viewed are the set of internal pages that the user consumes during each visit.
- Time is the time that has gone by since a user accesses the website and leaves.
- Content is each of the pages viewed on the website.
- Interesting KPIs are obtained by mixing the different metrics:
  - Percentage of recurrent users: are recurring users against new users, expressed as a percentage.
  - Average of pages visited per session: the higher this average, the higher the quality of content.
  - Average visit duration: the longer the average duration of the visit, the higher the quality of web content.
  - Bounce rate: this metric is linked to web quality and reflects users who visit a single page on the website and leave before 30 seconds of activity.

The analysis can be complicated further by using filters and segments.

Filters: Filters are there to focus on a few data. Filters are recommended to limit or modify the data in a view. For example, you can use filters to include only data from specific URLs or keywords, you can also exclude traffic from particular IP addresses or other types of information.

Segments: A segment is a subset of your analytics data. For example, of your entire user set, one segment might be users from a particular age range, or a particular country or city. The great advantage of using segments is that they allow to isolate and analyze those subsets of data, so you can examine and respond to the component trends in your business (Google, n.d.). For example, if in a subset of de Analytics data you find that users from a particular geographic region are no longer subscribing to your newsletter, you can offer some type of loyalty discount or a free download to foster subscription to the newsletter in that particular market.

There are three kinds of segments: subsets of users, subsets of sessions, and subset of hits.

Google Analytics allows you to use up to 100 segments.

6.4.1.1 e-Commerce Websites

Although e-commerce is measured using the usual web analytics, their specificity requires extra functionalities or hyper specific concepts that we need to manage, such as attribution modules.

In this chapter, we will discuss several ways of measuring websites, social media, and e-commerce—a multichannel approach through which each company or institution tries to connect with the end user. It is in this context that the need to generate new attribution models arises.

An attribution model is a set of rules assigning a particular value to the different channels a user has gone through before completing a conversion. The attribution model is, thus, a predefined pattern that helps establish the importance of the different channels and marketing actions, within a business, to achieve conversions on a website.

There are several reasons to justify the importance of having attribution models.

1. Once we know the true impact of each channel, we will be able to optimally distribute investment in advertising in the different media.
2. They can also clarify which campaigns in particular attract users in the first place (this is usually linked to campaigns that lead the user to conversion).
3. They are extremely useful to measure intangibles such as branding, as the impact of indirect profitability can easily be measured through branding campaigns, be it through social media or any other means.

6.4.2 Social Media: Main Metrics on Facebook, Twitter or Instagram

Social media is becoming one of the most widely used channels by companies in their marketing strategies; however, measuring the outcome of actions on these platforms might prove difficult, at least if we try to measure impact in terms of ROI. That is the reason why social media requires a discussion of IOR (Impact of Relationship; Alrubaiee & Al-Nazer, 2010), that is, the impact that social media has in the sales objectives.

Despite large differences across platforms, the levels of interaction on the different media are quite similar due to a homogenization process in which platforms introduce as improvement innovation coming from their competitors ("share" on Facebook is inspired on retweet, just like Twitter's like star is a heart like Instagram's). Therefore, Kaushik (2011) suggests a classification of metrics based on less to more interaction:

- Applause: Instagram heart, Facebook like, or Twitter fav, for example.
- Sharing: Facebook share, Twitter retweet, Pinteres repin, etc.
- Public conversation: Facebook comment, Twitter reply, or cited tweet, etc.
- Private conversation: private message on Facebook messenger, direct messages on Twitter, etc.

The higher the level of interaction (more effort or proactiveness by the user), the higher the importance this metric should be given. And yet, many interactions have their equivalents in different social networks (Table 6.1).

Sharing metrics are also known as "scope", or the volume of content dissemination on social media. Regarding scope, we have to make a difference between paid (stemming from advertising campaigns on a social platform such as Facebook ads) or unpaid scope. Within unpaid scope, there is an organic reach when the content reaches your contacts or the contacts of those you have tagged, while viral reach is estimated over the total people who have seen the post not through the original post, but through sharing and interactions by other users.

Another question that is sometimes neglected and that must be closely monitored are shared URLs through social media. A URL link shortener such as Bit.ly or similar resource is recommended to measure the performance of a publication (Figure 6.3), including links, such as the number of clicks or devices from which the URL has been accessed, as well as the traffic sources (the networks where it has been shared, for example). The outcomes can sometimes be peculiar, such as a tweet that has been retweeted more than the clicks on the link it contains.

Although web analytics offer sources of traffic—social traffic, in this case—shortened links produce concrete statistics on specific links, regardless of the destination. Another great advantage is that they allow the analyzer to monitor the performance

TABLE 6.1
Main Metrics and Equivalents on Social Media

Social Media	Applause	Sharing/ Amplification	Public Conversation	Private Conversation
Facebook	Like	Share	Comment	Private message, messenger
Instagram	Heart	Repost	Comment	Direct message
Twitter	Fav	Retweet	Reply, cite with a comment	Direct message
LinkedIn	Recommend	Share	Comment, share with comment	Private message

Source: Prepared by the authors.

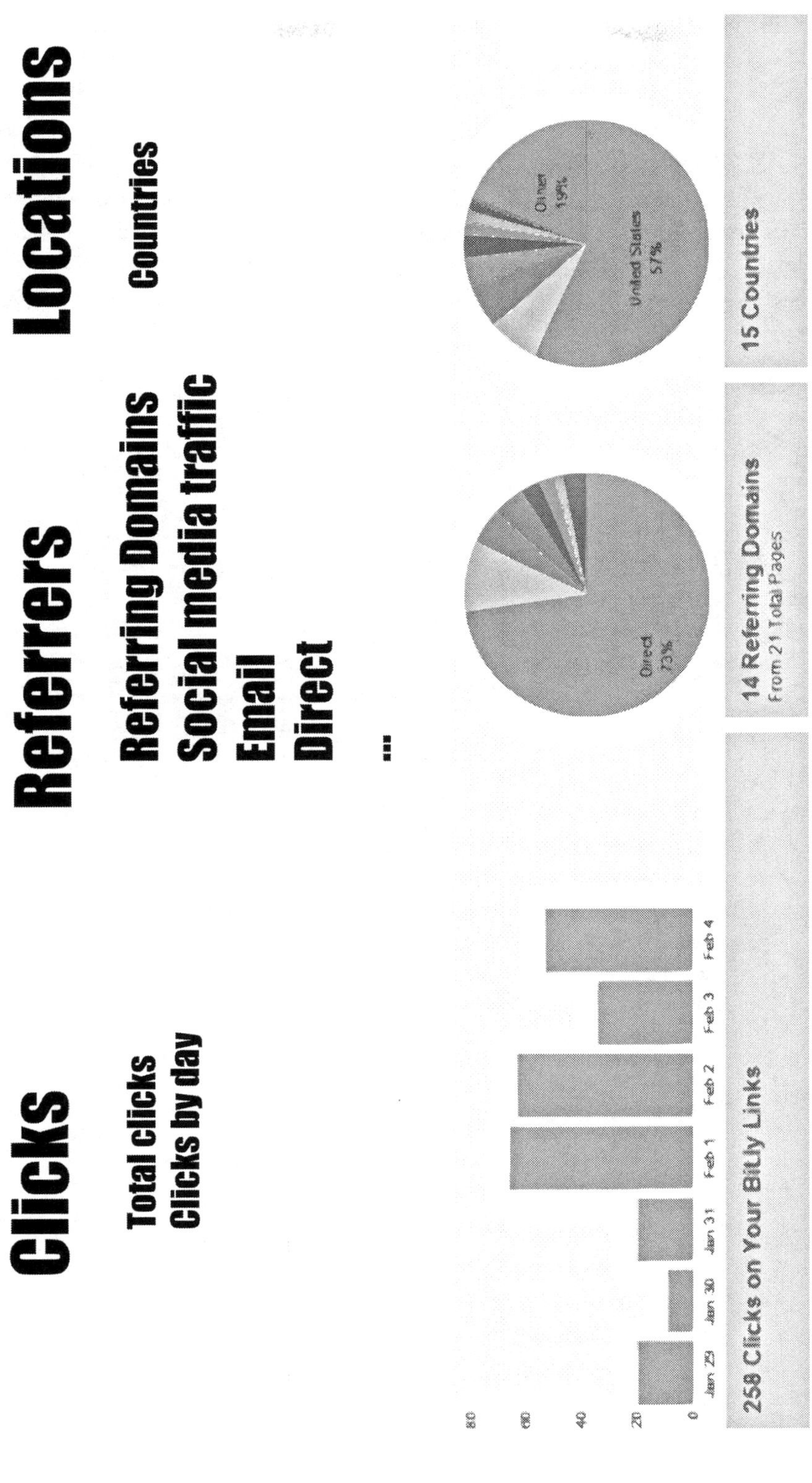

FIGURE 6.3 Advantages of shortened clicks. (Source: Prepared by the authors.)

of URLs shared by competitors as the analytics of shortened links is open (in bit.ly one has to just add a + at the end of the URL on the browser and click enter to see the performance of this link).

6.4.3 Newsletters

E-mail marketing is the act of sending a commercial message, typically to a group of people, using e-mail. Despite the passing of time, this is still a very efficient tool. E-mail predates the web. From the first e-mail in 1971 (Computing History), this technology's structure has remained stable with the fields "user", "subject", and "message". In fact, in order to register on any social media, one has to usually have an e-mail address.

As a digital marketing tool, e-mail marketing can also be assessed through different metrics, subdivided into families of indicators: the user database, sending the newsletter, and conversion metrics. All metrics can work together in the form of an email marketing conversion funnel (figure 6.4).

- The user database
  It includes the metrics related to databases. In particular, the subscriber increase rate and unsubscriber rate both express a percentage regarding a previous point in time in the database. Measuring an increase or decrease in a database or a user segment helps identify the causes behind the change. Besides, as these metrics are presented on a time axis, we can establish a cause-effect relationship to some content (an e-mail marketing campaign), so that we may get to know which content works better or worse (causing an increase or decrease in users).
- Sending a newsletter
  - We can use four indicators to assess the success of e-mail marketing:
    - Sending rate: This is the percentage of items correctly delivered to the recipient (i.e., that have not reported an error).
    - Opening rate: Percentage of e-mails that, after correct delivery, have been opened by the user. Some apps allow us to know what users have received and opened their e-mail, as well as the time when they received and opened them.

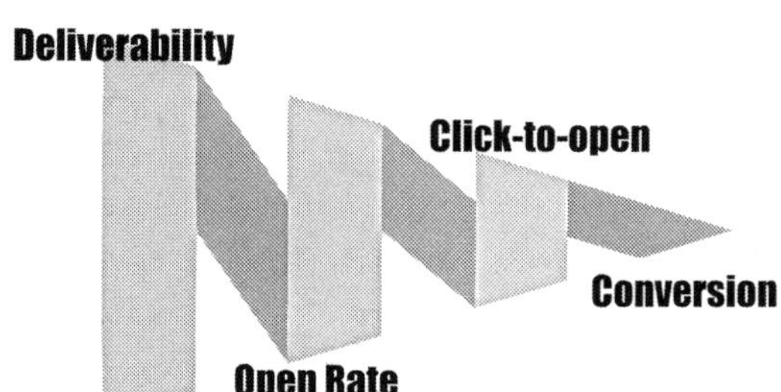

FIGURE 6.4 Funnel conversion measuring techniques for an e-mail marketing campaign. (Source: Prepared by the authors.)

   - Clicks on links: Within the correctly delivered and opened e-mails, it is possible to determine which links have been clicked on and the number of times those links have been accessed.
   - Unsubscriptions per delivery: Of the correct and opened e-mails, we can identify the number of people who have cancelled their subscription to a newsletter after a certain dispatch.
- Conversion metrics
  Apart from the database and the technical dispatching data, one can set conversion rates to check the direct relationship between the campaign and the accomplishment of particular goals. Some examples of e-mail marketing could be downloading a mobile app, using a voucher, registration on another website, sending information from the user, etc.

  An e-mail marketing strategy is perfectly compatible with social media marketing, and it often supports the development of apps/sites. We will not address SMS marketing in this chapter (sending commercial information through SMS) because assessing it can only be done through measuring access to URLs, if present.

6.4.4 Mobile Apps

Within digital marketing, the latest trend focuses on the business of mobile apps, which have experienced a sharp increase in recent years. Despite its recent beginnings (we can date it back to Apple's App Store launch in 2008), both Google (Android) and Apple (iOS) have been spearheading the constant growth in the number of apps in their app stores. The volume of apps is such that the efforts in the mobile ecosystem focus more on the development of positioning strategies (app store optimization or marketing ASO) so that users find and test their apps. Mobile app analytics offers a detailed study of generic app evolution data, such as the volume of downloads and usage time, as well as internal usage dynamics related to the behavior of each user with the mobile app.

Technological developments now offer push notifications, messages from each app installed that can be sent from the central app to all devices that have that app installed and that allow for instant communication.

Push notifications have a very high ROI, as they improve engagement and focus on a very concrete target. In particular, push notifications by fast food or delivery services (for example, Just Eat) have been very effective in increasing sales.

6.5 ANALYZING AND REPORTING: WHAT A WEB ANALYTICS AND SOCIAL MEDIA REPORT SHOULD ANALYZE

A strategy's success will be mostly determined by measurements and analyses. When discussing a digital strategy, we need to go back to the concept of Business Intelligence: "a data-driven DSS that combines data gathering, data storage, and knowledge management with analysis to provide input to the decision process" (Negash & Gray, 2008, p. 175).

Business intelligence is the ability to change data into information, and that information into knowledge, in a way that can be optimized for business decision-making processes. Within the different Business Intelligence products, we have today, the most interesting ones here are Dashboards: these are business control tools to monitor the objectives of a company through the most important KPIs, with all metrics in one place and at a glance. A Dashboard is intended to monitor KPIs, not so much to a detailed analysis of each of the metrics.

Selecting a good tool that brings together the KPIs we want to analyze is important as it helps complement and grade information, but this is not enough for Business Intelligence.

As Eckerson (2010, p. 39) claims, "Performance dashboards provide a layered information service that combines monitoring, analysis, and reporting—and increasingly prediction and visualization—in a single integrated environment. By integrating the functionality of most BI tools, performance dashboards meet 60 percent to 80 percent of the information needs of most casual users".

Each business must be in a situation to create its own dashboard. Any dashboard, regardless of the data they process, must comply with the following aspects:

- It must be totally customizable, so that it can adapt to all sorts of businesses and institutions.
- It must allow easy data visualization in line with current strategies of "Smart Visual Data".
- It must have different functionalities, such as the possibility to interact with external apps or services.

Some common tools to create customized dashboards can be found in the section Glossary of Tools.

6.6 WHERE SHOULD THE EFFORTS OF SMALL AND MEDIUM SIZE ENTERPRISES BE INVESTED

Most SMEs do not perform web traffic analysis nor analyses on the performance of their content on social media (if they even have any!). Most likely, they lack the resources to hire a full-time professional. That is why we have tried to tackle some of the reasons why SMEs should use them and focus on them.

In an environment in which lack of innovation in technology and marketing can be costly, data analysis can become a strong competitive edge in the sector.

Going through all the things we have mentioned in this chapter might be tough for marketing managers in SMEs—our recommendation is therefore to focus on a web analytics and social media strategy around the following aspects:

1. Design a feasible strategy and goals. The goals are the backbone of any strategy and as such they have to be the horizon to follow at all times, the North star to guide us in the digital storm. Design a strategy that can be measured with concrete and perfectly measurable KPIs.

2. Direct users to what you want them to do (leads). Establish the conversations you deem necessary and guide your communication and marketing strategy toward those specific leads.
3. Check that your design and strategy work. Remember that if you cannot measure it, you cannot asses it and you will not know whether your strategy is working. A good dashboard with basic KPIs will help you assess if the design and strategy work.
4. Iterate. This is the moment to come back to square one with a bag full of experiences after implemented and measured actions. Go through what was done well and not so well, redefine your goals if need be, or suggest a new business model or market segment after analyzing your data. Iteration is the key to survive in a liquid environment in constant transformation.

In a competitive and sometimes volatile environment, SMEs cannot be competitive without a digital marketing strategy. In this context, clear objectives are paramount, but also how to measure them—metrics and KPIs. We live in the data era, what and why to measure are still transcending questions. So much so that what Montague announced 200 years ago is still valid: what we cannot measure, doesn't exist. And if a company cannot measure its digital strategy, then it does not have one.

REFERENCES

Alrubaiee, L., & Al-Nazer, N. (2010). Investigate the impact of relationship marketing orientation on customer loyalty: The customer's perspective. *International Journal of Marketing Studies*, *2*(1), 155–174.

Computing History. (1971). First network email sent by Ray Tomlinson. Retrieved from http://www.computinghistory.org.uk/det/6116/First-e-mail-sent-by-Ray-Tomlinson/

Doran, G. T. (1981). There's a S.M.A.R.T. way to write management's goals and objectives. *Management Review*, *70*, 35–36.

Eckerson, W. W. (2010). *Performance dashboards: Measuring, monitoring, and managing your business*. Hoboken, NJ: John Wiley & Sons.

Google. (2019). About demographics and interests. Analytics help. Retrieved from https://support.google.com/analytics/answer/2799357?hl=en

Google. (n. d.). Fundamentos de Marketing Digital. Retrieved from https://learndigital.withgoogle.com/activate/course/digital-marketing

Holt, E. B., Marvin, W. T., Pepperrell Montague, W., Barton Perry, R., Pitkin, W. B., & Gleason Sapulding, E. (1912). *The new realism: Coöperative studies in philosophy*. New York: Macmillan.

Kaushik, A. (2011, October). Best Social Media Metrics: conversation, amplification, applause, economic value [online article]. Retrieved from https://www.kaushik.net/avinash/best-social-media-metrics-conversation-amplification-applause-economic-value/

Kelvin, W. T. B. (1883). *Electrical units of measurement: Being one of the series of lectures delivered at the Institution of Civil Engineers, Session 1882–83*. London: Institution of Civil Engineers.

Morell, J. M. C. (2015). *Zen Business: Los beneficios de aplicar la armonía en la empresa*. Barcelona: Profit Editorial.

Negash, S., & Gray, P. (2008). Business intelligence. In F. Burstein & C. W. Holsapple (Eds.), *Handbook on decision support systems 2* (pp. 175–193). Berlin, Heidelberg: Springer.

Sharma, H. (2016, July 23). Google analytics users (new, returning, unique) explained in great detail. Retrieved from https://www.optimizesmart.com/understanding-users-in-google-analytics/

WAA Standards Committee. (2008). *Web analytics definitions*. Washington DC: Author.

GLOSSARY OF TOOLS

Finally, we would like to offer a short list of reporting tools to monitor the most relevant web and social media metrics. The selection of tools is based on efficiency, that is, that they are affective at the lowest cost. Many tools are free or have a free version to track interesting data (and assess if it is worth getting the full app).

Chartio: This is a dashboard that allows for the integration of external databases in mysql. https://chartio.com/

Google Analytics: This is the most basic web analytics tool. It allows you to measure your audience, the behavior of users on your web, and conversions of visits to your web. https://analytics.google.com/analytics/web/

Hootsuite: This is one of the most widely used tools in the world for multi-account social media management, but it is also useful to prepare social media analytics reports. A competitive advantage against other tools of this kind is that it analyses shortened links as it has its own shortener ow.ly. You can monitor the performance (in clicks) of the links shared on social media. https://hootsuite.com/

In-house analytics: Facebook fan pages and Instagram business accounts have their own social media analytics tools. In particular, Instagram's tool is very comprehensive, while Facebook's is complex to use due to the immense number of metrics it provides.

MeltWater Likealizer: This is a tool to obtain clear and simple metrics of the fan pages it manages, with interesting metrics such as total followers, likes growth, engagement, and ranking. Besides, it also offers tips and similar sites to improve your ratings. You can also look for other fan pages and take their metrics as a benchmark. https://likealyzer.com/

Pirendo: Is a social media analytics tool with a complete dashboard to monitor all networks, content, passwords, etc. https://pirendo.com/

SumAll: Is a tool that offers multiplatform social media analytics about a large number of social services, from the main social networks (Twitter, Instagram, Facebook pages, YouTube, LinkedIn …) to e-commerce platforms or payment gateways (shopify, eBay, PayPal, stripe, …) so that a single platform can help monitor one user's accounts. https://sumall.com/

Twittonomy: This is a very powerful tool to monitor Twitter accounts. It allows you to monitor users, lists, and keywords in a simple and highly accurate way. It is limited by Twitter's API (application programming interface), with access only for the last 3200 tweets for each search. http://www.twitonomy.com

Visually: Is a dashboard for data visualization. In fact, infographics and presentations may be produced from the tool itself. https://visual.ly/

7 Managers' Perception of Business Intelligence Capability of SMEs in Turkey

Mine Afacan Findikli
Department of Business Management,
Beykent University, Istanbul, Turkey

Mustafa Sundu
Department of Business Management,
Beykent University, Istanbul, Turkey

Okan Yaşar
Beykent University, Istanbul, Turkey

CONTENTS

7.1 INTRODUCTION

In today's world, two most prominent features of business organizations are "uncertainty" and "complexity" (Mowles, 2015; Renn, 2017). The continuous development of information and data, and the increasing environmental dynamism cause pressure on organizations' activities (Schick, Frolick, & Ariyachandra, 2011). Thus, the biggest issue that forces business organizations in such an environment is to be able to realize accurate and timely decision-making for the organization's sustainability.

At the same time, the assets of business organizations have transformed from material assets to intangible assets, and they are now governed by intellectual capital and knowledge-based practices (Olszak, 2016, p. 105). Whereas the production and the trade are managed by "knowledge", the ability to benefit from all available information and knowledge is as a crucial issue, and it requires a special competence to obtain the right information as well as ensure its proper use. Business intelligence (BI) is one of the technologies that may provide such information.

Although there are successful examples of the use of BI, in Turkey, numerous small- and medium-sized enterprises (SMEs) that struggle to maintain their business cannot effectively use BI. For example, some of the SMEs tried to recruit information and communication technology (ICT) specialists and approximately half of them reported difficulties. As it can be seen in Table 7.1, 10.3% of SMEs have made an effort to recruit ICT specialists and 34.4% of these enterprises faced difficulties in 2017 (TurkStat, 2018).

Extensive research has been performed in the past two decades to understand the BI system mainly on three sections: adoption, utilization, and success of BI in organizations (Ul-Ain, Vaia, & DeLone, 2019). But, in Turkey, BI has not yet earned the place it deserves in the decision-making process and gaining competitive advantage within the business organizations. Moreover, the antecedents of failures in this area have not yet been fully examined. For this reason, it is especially important for organizations to conduct BI research systematically (Olszak, 2016, p. 111).

Based on this perspective, our research questions are as follows:

1. What is the extent of BI adoption in Turkish SMEs?
2. What are the challenges and opportunities that BI impose to SMEs?
3. What is the BI capability of Turkish SMEs?

To answer these questions this chapter is organized as follows: The first part provides the theoretical foundations for the BI development. The literature was drawn

TABLE 7.1
SMEs that Recruit or Try to Recruit ICT Specialists and Reported Hard to Fill Vacancies

Size/Class	Period	Enterprises that Recruit or Try to Recruit ICT Specialists (%)	Enterprises which Recruited or Try to Recruit ICT Specialists Reported Hard to Fill Vacancies (%)
10–49	2013	3.9	39.0
	2014	5.4	58.6
	2015	3.9	41.5
	2016	3.0	40.4
	2017	3.7	39.4
50–249	2013	9.7	28.7
	2014	13.8	40.6
	2015	9.3	39.9
	2016	8.4	33.5
	2017	10.3	34.4

to identify the BI, need for BI, opportunities and challenges of BI, BI adoption of SMEs, importance and future of BI, and BI capabilities. And then an empirical research is conducted and reported.

7.2 NEED FOR BUSINESS INTELLIGENCE

With the globalization phenomenon, an economic development, new trade regulations, or an innovative product at one end of the world can impact the economy of another country. Companies must be much faster to take more accurate decisions in order to catch up with updated standards. When all institutions work in such a competitive environment, a risk factor arises. As a result, the need for BI has expanded with an increase in the number of decisions to be made per unit time. In this respect, the key to sustainable success in competition is to respond properly and quickly to new market conditions while regarding consumer expectations.

Due to the fact that any information acquired and used is important in creating value, business organizations use BI applications to learn more about broader markets, to improve internal process performance and to progress over time. In addition to being vital to a brand and product, the data flow at every step in the value chain is a source for potential new initiatives and solutions to problems. Besides, to achieve their goals with high performance, they understand the benefits of regularly monitoring and analyzing the data flow and acquiring and sharing new information with shareholders. Thus, they take advantage of numerous and continuous data flows acquired from operational transactions, sales channels, and feedback in order to predict market changes, consumer expectations, and competitors' attitudes.

For the first time in history, the BI concept was used in an article written by an IBM researcher Hans Peter Luhn in 1958. He (Luhn, 1958, pp. 314–316) defined BI as a collection of various architectures and technologies that convert data into valuable information for business purposes to achieve a desired goal. BI is also explained as a broad range of applications, technologies and processes for collecting, storing, accessing, and analyzing operational data to provide timely competitive information for better insight into operational and strategic decision-making processes (Negash & Gray, 2008). Additionally, BI encompasses data warehousing as well (Krawatzeck & Dinter, 2015, p. 179).

So, BI is not just about the past or today's picture. It also serves to model the future of the businesses via predictive analytics. The current literature highlights the significant benefits of using BI in five categories (Thomas, 2001, p. 47).

- **Artificial Intelligence (AI) category** includes AI algorithms and applications.
- **The benefits category** is described by the organizations on how to use data collection, data mining or enterprise-level BI systems to achieve measurable financial benefit.
- **The decision category** includes articles on improving general decision-making topics such as data modeling, decision-making, and decision modeling.

- **The application category** includes project management, including data storage, data mining, customer relationship management (CRM), enterprise resource planning (ERP), and knowledge management systems (KMS).
- **Strategy category** deals with application of BI tools in business environment. The category includes improving internal performance such as business agility, interdisciplinary integration, collaboration with external partners and improvement of the supply chain, and CRM.

Analyses carried out to monitor the market trends of BI solutions show that the entire market grows by 7.9% annually and the business-led solution, with the rate of 63.6% in 2015, is in rapid growth, and the decision to invest in BI systems has been greatly influenced by the increased need for flexibility in use and the personalization of data and information (Caserio & Trucco, 2018, p. 45).

Considering these points, we can conclude that BI supports decision makers with decision support systems (DSS), appropriate software technology products, and analytics in terms of the strategic level decision-making process and data driven business (Olszak, 2016; Visinescu, Jones, & Sidorova, 2017). Findings from the literature show the positive impact and added value of BI systems on sales, marketing, and inventory management operational processes (Elbashir, Collier, & Davern, 2008, p. 149). It is shown that the BI systems provide accurate and timely information about customer needs in many different segments and support managers to better match products to market demand (Long, Shelhamer, & Darrell, 2015).

In conclusion, based on institutional theory, it is suggested that companies are forced to change by their environmental factors (DiMaggio & Powell, 1983; Scott, 1995). Thus, the company's decision to invest in BI has become evident by three main drivers (Caserio & Trucco, 2018, p. 46).

The first driver of BI needs is stated as coercive isomorphism that is related to the need to manage a growing amount of data and information. As is known, the spread of computer technology and communication tools has greatly increased the amount of data and information. In this data stream, business managers should run against seconds to make rapid evaluations and to respond correctly. Faced with an increasingly complex environment and more data, and the necessity of compliance with regulations, businesses are forced to act quickly (Haveman, 1993). According to the coercive isomorphism, business organizations, which are found to be dependent on their stakeholders, change their institutional practices because of the pressure from the social, governmental, and technological changes (Powell & DiMaggio, 2012). The second drivers for BI need are expressed as mimetic isomorphism. Mimetic isomorphism leads businesses to search for imitating information technology practices of rivals. Because of the coercive pressures from the regulatory bodies, investors, and software providers, they feel the pressure to imitate the use of BI (Kaya & Akbulut, 2018). The third drivers for BI need are stated as decision-making process, associated with the need to deal with more complex processes due to highly competitive contexts that require the increasing use of advanced information technologies and sophisticated decision-making algorithms (Turban, Sharda, & Delen, 2014).

To manage these requirements and pressures, companies need to adapt BI for several business functions and areas. The main areas where BI is needed are management information systems, strategic planning, marketing, regulations, and financial fraud detection (Caserio & Trucco, 2018). Management information systems can be used to coordinate and integrate organizational and technical issues (Hou, 2012, p. 568) for improving data management and internal and external interaction (Peters, Wieder, Sutton, & Wakefield, 2016, p. 7). In strategic planning, BI is used to monitor signals coming from different environments (Alkhafaji, 2011). Besides, planning and monitoring the requirements are essential areas for firms. In marketing, BI is used to understand customer needs, to monitor their loyalty (Olszak, 2016, p. 108), and to implement marketing strategies (He, Zha, & Li, 2013).

BI is a part of the strategies and technologies that enterprises utilize for the data analysis of business information in order to obtain historical, current and predictive insights to increase the value of business operations. Although it is an effective tool that can be used to transform the data into efficient and meaningful information, it is only as valuable as the business outcomes.

According to the literature, the measure of BI functions encompasses two main fields (Lönnqvist & Pirttimäki, 2006, p. 468). The first one is to question its effectiveness for the firms (Sawka, 2000, p. 54). Managers should regularly controlindicators to confirm the department's service quality (Davison, 2001, p. 27). Second, they should measure BI processes, because BI outputs should be rational and satisfy the stakeholders (Lönnqvist & Pirttimäki, 2006).

The findings about the value of BI answer two main questions about the cost of BI and utility of BI. On the other hand, experiences indicate that the achievement of BI is still controversial. While putting into practice, a great deal of BI applications fail. Firms do not receive intended results from the use of these applications (Chaudhary, 2004; Howson, 2007; Isik, Jones, & Sidorova, 2011, p. 168; Schick et al., 2011). Findings report that about 60–70% of applications have no success, because of technological, cultural, organizational, and infrastructural factors (Clavier, Lotriet, & Loggerenberger, 2012, p. 4144). Thus, it is understood that most organizations seem to be unable to use the BI tools for managerial decisions as a competitive advantage.

The foremost critical issues that whether BI is effective within the organizations or not are related with BI and its utilization such as quality of information, abilities, sponsorship, and the arrangement between BI and the organization (Davenport & Harris, 2007). Other components are related to the organization itself such as organizational culture, data necessities, and legislative issues. One of the greatest boundaries that the organizations experience during the execution of BI frameworks is to have a business and organizational character (Olszak & Ziemba, 2012, p. 135). The key organizational boundaries can be summarized as follows: the need of manager's support and information about the BI framework and its capabilities, the budgeting of the BI, complicated BI venture, and the need for client training and support (Olszak, 2016, p. 108).

7.3 THE FUTURE OF BUSINESS INTELLIGENCE

According to Digital World Survey conducted by IDC (International Data Corporation), it is estimated that only 0.5% of the data is analyzed all around the world. Data will grow 40% by 2020 and analysis capacity could remain insufficient unless technological investment is done accordingly (Gantz & Reinsel, 2012). Based on the McKinsey Global Report, 90% of the data in today's world was created in last two years, and the report emphasizes that by 2020 the data size will be 44 times bigger than it was in 2011 (Gobble, 2013). Moreover, Rubinstein (2013) stated that Big Data has some resemblance with the tsunami that urges the need to take measures against this. It will become more important to analyze than to collect data. Therefore, it is obvious that the importance of BI practices will increase even more. By so far, BI has been involved in data-centered management (Chaudhuri, Dayal, & Narasayya, 2011, p. 95). In the long term, BI will be significant for everybody; not only for data specialists and internal employees, but also for all stakeholders.

Chen and his colleagues (2012) summarize a broad overview of BI in terms of evolution, application, and emerging trends. They stated that the term BI 1.0 is the BI technology and applications currently in use in the industry. In this area, structured data is often used; the same data is collected by older generation systems and is usually stored in relational database management systems (RDBMS). BI 2.0 research includes web intelligence and web analytics, consisting of data collected by web 2.0-based social media and web content. BI 3.0 will contain integrated Big Data analytics based on mobile applications and sensors (Chen, Chiang, & Storey, 2012).

7.4 THE CHALLENGES FOR BUSINESS INTELLIGENCE PRACTITIONERS

Pierre Nanterme, CEO of Accenture, stated that almost half of the companies went out of business since the year 2000 due to rapid developments in digitalization, and clarified his statement that digital forces have a great impact on businesses. It is noticed that under these changes businesses try to regulate their business only by digitalizing their channels in the front office, addressing marketing, sales, and services. However, it is obvious that this is not a sustainable strategy (Nanterme, 2019). In other words, businesses should properly understand and behave BI and digitalization to achieve sustainable growth. However, SMEs managers or entrepreneurs may tend to continue to work with conventional practices due to their working culture, or as Nanterme has already declared, they try to regulate their process superficially due to the lack of understanding. The most important thing is not only the acquisition of the data, but the competence for transformation of data into information. Moreover, qualified workforce with digital literacy is needed to use this information to predict and shape the future for business.

For this reason, business executives should first know that they would face a significant cost to implement a robust BI framework and to sustain it. Second, it is necessary to recruit qualified and engaged staffs/teams to design, implement, and support this framework (Gash, Ariyachandra, & Frolick, 2011, p. 263). But challenges facing BI practitioners are not only limited by these two issues. According to O'Donnell and his colleagues (2012, p. 207), BI practitioners in Australia have faced 20 challenging issues about BI, such as developing a BI strategy, data management, engaging and training users, and security. First, they have identified that practitioners were very interested in comprehending processes for increasing data quality. However, among a large number of different techniques and technologies, they were feeling a little bit confused in terms of choosing the appropriate data management tools. Second, they have noticed that the practitioners have wanted to understand the best strategies and tools to use predictive analytics. Third, even though they didn't have any idea how to structure an efficient BI team, they have agreed on its necessity. For this reason, the researches have stressed the value of engaging and training of BI practitioners about the value and outcomes of BI.

From SMEs perspective there are also significant issues to mention. First, the implementation and adoption of BI could be more difficult because of the operating cost of deploying BI. It is already known that budgets and resources are limited especially for SMEs. In addition to the cost of implementation, setting up these data warehouses and processors also force to find IT labor resources. Therefore, today, BI projects are relatively large-scale projects when they are examined in terms of cost and need and are preferred by relatively large corporations (Seker, 2016).

Second, in a small business, a data culture may not exist, and departments may be discouraged from a lack of time, data resources, and the adoption of BI. They may not see the benefits over the adoption costs. Olszak (2016) reported that about 60–70% of BI applications fail, due to the technology, organizational, cultural, and infrastructure issues. According to a study by Garcia and Pinzon (2017), today's BI integrations fail 70–80% due to technological and managerial reasons. In light of these findings, the related researches suggest that BI does not consistently live up to those expectations; in particular, shortcomings in BI's planning, establishment, employee skills, technology, and strategy (Visinescu, Jones, & Sidorova, 2017). In this respect, we can argue that SMEs have faced many challenges to benefit from BI as it should be.

7.5 SME AND BI USAGE IN TURKEY

The definition of SME in Turkey suggests certain differences because of legal frame built before 2005 that has caused many problems for SMEs. For example, an SME could benefit from KOSGEB (Small and Medium Enterprises Development Organization of Turkey), but it could not benefit from Exim bank's export credit. To eliminate the conceptual confusion, in 1996, the European Union developed an SME definition based on three criteria: number of employees, balance sheet size, and independence (Table 7.2) (European Commission, 2015, p. 11).

TABLE 7.2
EU SME Criteria

Enterprise Category	Headcount: Annual Work Unit (AWU)	Annual Turnover (€)	Annual Balance Sheet Total
Micro	≤10	≤2 million	≤2 million
Small	≤50	≤10 million	≤10 million
Medium size	≤250	≤50 million	≤43 million

TABLE 7.3
Turkey SME Criteria

Enterprise Category	Headcount: Annual Work Unit (AWU)	Annual Turnover (₤)	Annual Balance Sheet Total
Micro	≤10	≤1 million	≤1 million
Small	≤50	≤8 million	≤8 million
Medium size	≤250	≤40 million	≤40 million

In 2005, the Council of Ministers of Turkish Republic prepared a regulation on "Classification, Qualifications and Definition of SMEs" to define SME. "Enterprises whose 25% or more of their capital or shares are not undertaken by an enterprise or whose capital does not consist of enterprises other than SME" are enlisted as SME. The criteria specified in the regulation are presented in Table 7.3 (Implementing Regulation on the Definition, Qualifications and Classification of SMEs, 2012; KOSGEB, 2015).

Since the economy is largely based on SMEs, which form 99.8% of all Turkish enterprises, SMEs are crucial economic factor for Turkey. Recent reports show that 30% of Turkey's exports are conducted by SMEs (European Commission, 2019). In order to understand the importance of the role of SMEs in the Turkish economy, Table 7.4 provides a significant statistics (European Commission, 2018).

In terms of BI usage in SMEs in Turkey, we can notice that BI-related statistical studies are concentrated in two areas: BI infrastructure such as computer use, Internet access, and software; ERP, CRM, and Supply Chain Management (SCM) usage (TurkStat, 2018).

Analysis by years show that computer use and internet access rates increased between 2010 and 2018. As of 2018, the number of SMEs with 10–49 employees is 96.6% and Internet access is 94.7%. SMEs with 50–249 employees have 99% computer access and 97.8% Internet access (Table 7.5).

TABLE 7.4
Turkish and EU SMEs Comparison and Shares in Economy

	Number of Enterprises			Number of Persons Employed			Value Added		
	Turkey		EU28	Turkey		EU28	Turkey		EU28
Size Class	Number	Share (%)	Share (%)	Number	Share (%)	Share (%)	Million (€)	Share (%)	Share (%)
Micro	2.380.885	96.4	92.8	5.112.590	39.2	29.0	30.075	15.3	20.2
Small	57.214	2.3	6.0	1.754.015	13.5	20.2	24.717	12.6	17.7
Medium	25.948	1.1	1.0	2.582.542	19.8	17.1	44.596	22.6	18.4
Total	2.464.047	99.8	99.8	9.449.147	72.5	66.3	99.388	50.5	56.3

TABLE 7.5
SME Internet Access and Computer Usage

Size Class	10–49		50–249	
Reference Period	Computer (%)	Internet Access (%)	Computer (%)	Internet Access (%)
2010	91.3	89.7	97.0	96.9
2011	93.0	91.4	98.1	96.7
2012	92.5	91.2	98.2	98.1
2013	90.7	89.3	97.5	97.0
2014	93.5	88.5	98.3	96.1
2015	94.3	91.1	98.7	98.0
2016	95.3	92.8	98.2	96.9
2017	96.9	95.4	98.5	97.8
2018	96.6	94.7	99.0	97.8

With regard to software usage (ERP, CRM, and SCM), it is seen that there is a decrease in ERP and SCM usage but an increase in CRM between 2012 and 2017. As of 2017, SMEs with 10–49 employees have an ERP usage rate of 10%, CRM usage rate of 16.9%, and SCM usage rate of 8.1%. SMEs with 50–249 employees have an ERP usage rate of 26.1%, CRM usage rate of 24.5%, and SCM usage rate of 11.6% (Table 7.6).

TABLE 7.6
Enterprises Using Enterprise Resource Planning (ERP), Customer Relationship Management (CRM), and Supply Chain Management (SCM) Software

Size Class	Period	ERP (%)	CRM (%)	SCM (%)
10–49	2012	14.1	8.0	16.2
	2013	14.9	7.4	-
	2014	10.8	6.1	-
	2015	16.0	7.4	-
	2016	-	-	-
	2017	10.0	16.9	8.1
50–249	2012	32.6	13.5	23.2
	2013	31.9	12.7	-
	2014	25.7	11.7	-
	2015	32.6	14.5	-
	2016	-	-	-
	2017	26.1	24.5	11.6

According to TURSTAT statistics and based on the classification of Chen et al (2012), SMEs may be inferred to be in the B 2.0 level because of limited number of companies using related software and applications such as web intelligence and web analytics.

7.5.1 Research on Business Intelligence Adoption of SMEs in Turkey

This study aimed to measure the managers' perceptions of "Benefits and Challenges of Business Intelligence Adoption" in SMEs in Turkey. For this purpose, the perception scale developed by Scholz, Schieder, Kurze, Gluchowski, and Böhringer (2010) is used. The scale has two subdimensions, namely the benefits (18 items) and challenges (19 items) of BI. The sample of research consists of 126 Turkish SMEs that were selected randomly. Statistics for research sample is presented in Table 7.7.

Five Likert scale was used to measure managers' perceptions of the benefits and challenges for BI systems used in SMEs. In addition, ANOVA analysis was performed to find out whether managers' perception of adoption of BI differs in terms of operation field, sector, size, and income level. BI perception of SMEs statistics is given in the Table 7.8.

According to Table 7.8, managers' perception scores about benefit of BI are higher than perception scores about challenge of BI in terms of size, sector, operation field, and income level. For determining the significance of differences of perception score, Tukey test is conducted. According to the analysis, it was found that the perception of the managers does not differ in terms of the firm's income level ($p = 0.604$). For this reason, Tukey test outputs of this variable are not included in the Table 7.8.

The perception of benefit differed significantly in terms of micro- (1–9 employees) and medium-sized (50–249 employees) SMEs ($p = 0.001$). Similarly,

TABLE 7.7
Research Sample (126 SMEs) Statistics

Operation Field	National	54
	International/Global	23
	Regional	49
Sector	Manufacturing	34
	Service	39
	Trade	53
Size	1–9 Employees	39
	10–49 Employees	41
	50–249 Employees	46
Income Group (₺)	2,000,000–9,999,000 income	24
	10,000,000–49,999,000 income	43
	50,000,000–99,999,000 income	39
	100,000,000–499,999,000 income	17
	500,000,000 and over income	3

TABLE 7.8
SME Business Intelligence Adoption

SMEs Features	Challenges Mean	Challenges Std. Error	Benefits Mean	Benefits Std. Error
National	2.6105	0.75089	3.6148	0.33032
International/Global	2.3041	0.63091	3.9877	0.35002
Regional	2.7632	0.66823	4.0486	0.50832
2,000,000–9,999,000 income	2.9123	0.27784	3.5278	0.76719
10,000,000–49,999,000 income	2,4421	0.88587	4.0000	0.51370
50,000,000–99,999,000 income	2,6491	0.72821	3.9021	0.31569
100,000,000–499,999,000 income	2,5088	0.88279	4.0000	0.24216
500,000,000 and over income	2,2526	0.69166	3.9556	0.49752
1–9 Employees	2,6105	0.75089	3.6148	0.33032
10–49 Employees	2,3041	0.63091	3.9877	0.35002
50–249 Employees	2,7632	0.66823	4.0486	0.50832
Manufacturing	2,8596	0.63616	3.5370	0.38642
Service	2,5211	0.61049	3.8944	0.42909
Trade	2,4000	0.76988	4.1926	0.26962

perception of benefit differs significantly for sector variables. In addition to this, the perception of the managers about challenges of BI differs significantly in terms of the firm's operation field ($p = 0.016$). Table 7.9 summarizes the results of the Tukey test that show the significant differences between different groups.

It is seen that firms operating in the national field perceive challenges in BI applications compared to global companies operating in the international field (Table 7.9). However, in terms of the size, medium-sized firms (50–249 people) perceive much more benefits in BI compared to micro-sized firms (1–9 people). Trade SMEs perceive much more benefit in BI compared to manufacturing SMEs.

TABLE 7.9
ANOVA Analysis Outputs

Dependent Variables			Mean Difference	Std. Error	Sig.	Lower Bound	Upper Bound
Challenges	National	Regional	0.75439	0.31964	0.060	−0.0260	1.5348
		International/Global	**0.83219***	0.21384	**0.001**	0.3101	1.3543
Benefits	1–9 employees	10–49 employees	−0.37284	0.17513	0.098	−0.8004	0.0547
		50–249 employees	**−0.43380***	0.14928	**0.017**	−0.7983	−0.0693
	Manufacturing	Service	−0.3574	0.1465	0.0502	−0.7151	0.0002
		Trade	**−0.65556***	0.13104	**0.000**	−0.9755	−0.3356

*The mean difference is significant at the 0.05 level.

7.6 CONCLUSION AND DISCUSSION

Nowadays, for any enterprise, regardless of its size, sector, or capital stock, the core issue is to be effective and efficient to compete and survive. To achieve this goal, it is a primary management responsibility to adapt method, technology, and application. The existing human resources, organizational processes, customer relations, and supplier relations should be continuously improved in order to follow and adapt to changes in the environment. Especially disruptive technologies have transformed society as well as the management process and businesses throughout history. In this respect, management information systems have emerged with the adaptation of computer hardware and software technologies to operations in order to increase the efficiency of management. As computer hardware and software technology evolved, artificial intelligence applications began to analyze large data and thus, analytical forecasting capabilities of BI software applications have increased. Today, information technologies are the most important tool supporting management process, but it is not generally realizable for SMEs. The lack of capital and qualified human resources and appropriate technology may be prominent reasons for unsuccessful BI adoption.

In this respect, this chapter aimed to measure the managers' perceptions of "Benefits and Challenges of Business Intelligence Adoption" in SMEs in Turkey. Following the literature review, a research with 126 SMEs has been conducted. The research revealed that managers' perception about BI benefit and challenges differ in terms of characteristics of SMEs. The SME managers' perception of benefits was higher than the challenges for all characteristics of SMEs. Especially in terms of the number of employees, medium-sized enterprises (50–249) have a higher perception of benefit than micro-enterprises. Similarly, the benefit perception of trade SMEs was higher than that of manufacturing SMEs. The reason of benefit perception may be easy access to resources such as capital, IT expertise, and business network.

Meanwhile, the firms operating in the national field perceive more challenges in BI applications compared to global companies operating in the international field. Because globalization forces international SMEs to react instantly and to improve BI ability to be part of global operations and supply chains. One of the reasons why national SMEs perceive challenges could be information overload and insufficient understanding—lack of knowledge of exactly what it involves, where to start, and how long it will take for benefit—about BI.

REFERENCES

Alkhafaji, A. F. (2011). Strategic management: Formulation, implementation, and control in a dynamic environment. Development and Learning in Organizations: An International Journal, 25(2).

Arnott, D. (2004). Decision support system evolution: Framework, case study, and research agenda. European Journal of Information Systems, 13(4), 247–259.

Caserio, C., & Trucco, S. (2018). Enterprise resource planning and business intelligence systems for information quality: An empirical analysis in the Italian setting, 1st ed., Rome, Springer.

Chaudhary, S. (2004). Management factors for strategic BI success. In: M. S. Raisinghani (Ed.), Business intelligence in digital economy. Opportunities, limitations and risks (pp. 191–206). Hershey, PA: IGI Global.

Chaudhuri, S., Dayal, U., & Narasayya, V. (2011). An overview of business intelligence technology. Communications of the ACM, 2011, 54(8), 88–98.

Chen, H., Chiang, R. H., & Storey, V. C. (2012). Business intelligence and analytics: From big data to big impact. MIS Quarterly, 36(4), 1165–1188.

Clavier, P. R., Lotriet, H., & Loggerenberger, J. (2012). Business Intelligence challenges in the context of goods- and service-domain logic. Paper presented at the IEEE Computer Society 45th Hawaii International Conference on System Science, Maui, Hawaii, pp. 4138–4147.

Davenport, T. H., & Harris, J. G. (2017). Competing on analytics. The new science on winning. Boston, MA: Harvard Business School Press.

Davison, L. (2001). Measuring competitive intelligence effectiveness: Insights from the advertising industry. Competitive Intelligence Review: Published in Cooperation with the Society of Competitive Intelligence Professionals, 12(4), 25–38.

DiMaggio, P. J., & Powell, W. W. (1983). The iron cage revisited: Institutional isomorphism and collective rationality in organizational fields. American Sociological Review, 48(2), 147–160.

Elbashir, M. Z., Collier, P. A., & Davern, M. J. (2008). Measuring the effects of business intelligence systems: The relationship between business process and organizational performance. International Journal of Accounting Information System, 9(3), 135–153.

European Commission (2015). The SME definition. In: Publications Office of the European Union. https://doi.org/10.2873/782201

European Commission (2018). 2018 SBA Fact Sheet—Portugal.

European Commission (2019). Annual Report on European SMEs 2017–2018. SMEs growing beyond borders. https://doi.org/10.2873/248745

Gantz, J., & Reinsel, D. (2012). The digital universe in 2020: Big data, bigger digital shadows, and biggest growth in the far east. *IDC iView: IDC Analyze the future, 2007*(2012), 1–16.

Gash, D., Ariyachandra, T., & Frolick, M. (2011). Looking to the clouds for business intelligence. Journal of Internet Commerce, 10(4), 261–269.

Gobble, M. A. M. (2013). Big data: The next big thing in innovation. Research-Technology Management, January-February: 64–66.

Haveman, H. A. (1993). Follow the leader: Mimetic isomorphism and entry into new markets. Administrative Science Quarterly, 593–627.

He, W., Zha S., Li L. (2013). Social media competitive analysis and text mining: A case study in the pizza industry. International Journal of Information Management, 33(3), 464–472.

Hou, C-K. (2012). Examining the effect of user satisfaction on system usage and individual performance with business intelligence systems: An empirical study of Taiwan's electronics industry. Int J Inf Manag, 32, 560–573.

Howson, C. (2013). Successful business intelligence. New York, USA, Tata McGraw-Hill Education.

Kaya, I., & Akbulut, D. H. (2018). Big Data Analytics in Financial Reporting and Accounting, PressAcademia Procedia (PAP), V.7, pp. 256–259.

Implementing Regulation on the Definition, Qualifications and Classification of SMEs., Pub. L. No. Turkey/2005/9617 (2012).

Isik, O., Jones, M. C., & Sidorova, A. (2011). Business intelligence (BI) success and the role of BI capabilities. Intelligent Systems in Accounting, Finance and Management, 18(4), 161–176.

KOSGEB. (2015). 2015-2018 SME Strategy Plan. Retrieved from http://www.sp.gov.tr/upload/xSPTemelBelge/files/IcTlw+KSEP_Dokumani_2015-2018_10_08_15_YPK_ONAYLI.pdf

Krawatzeck, R., & Dinter, B. (2015). Agile business intelligence: Collection and classification of agile business intelligence actions by means of a catalog and a selection guide. Information Systems Management, 32(3), 177–191.

Long J., Shelhamer E., Darrell T. (2015). Proceedings of the IEEE Conference on Computer Vision and Pattern Recognition. Fully convolutional networks for semantic segmentation, pp. 3431–3440.

Lönnqvist, A., & Pirttimäki, V. (2006). The measurement of business intelligence. Information Systems Management, 23(1), 32–40.

Luhn, H. P. (1958). A business intelligence system. IBM Journal of Research and Development, 2(4), 314–319.

Mowles, C. (2015). Managing in uncertainty: Complexity and the paradoxes of everyday organizational life. UK, London, Routledge.

Nanterme, P. (2019). The World of Digital – Tholons. Retrieved July 13, 2019, from http://www.tholons.com/theworldofdigital/

Negash, S., & Gray, P. (2008). Business intelligence. In: Handbook on decision support systems 2 (pp. 175–193). Berlin, Heidelberg: Springer.

O'Donnell, P., Sipsma, S., & Watt, C. (2012). The critical issues facing business intelligence practitioners. Journal of Decision Systems, 21(3), 203–216.

O'Reilly, T. (2005). What Is Web 2.0? Design Patterns and Business Models for the Next Generation of Software, Retrieved Sep 30, 2019, from http://www.oreillynet.com/pub/a/oreilly/tim/news/2005/09/30/what-is-web-20.html

Olszak, C. M. (2016). Toward better understanding and use of business intelligence in organizations. Information Systems Management, 33(2), 105–123.

Olszak, C. M., & Ziemba, E. (2012). Critical success factors for implementing business intelligence systems in small and medium enterprises on the example of upper Silesia, Poland. Interdisciplinary Journal of Information, Knowledge, and Management, 7(2), 129–150.

Peters, M. D., Wieder, B., Sutton, S. G., & Wakefield, J. (2016). Business intelligence systems use in performance measurement capabilities: Implications for enhanced competitive advantage. International Journal of Accounting Information Systems, 21, 1–17.

Peters, T., Işık, Ö., Tona, O., Popovič, A. (2016). How system quality influences mobile BI use: The mediating role of engagement. International Journal of Information Management, 36, 773–783.

Powell, W. W., & DiMaggio, P. J. (eds.) (2012). The new institutionalism in organizational analysis. USA, Chicago, University of Chicago Press.

Renn, O. (2017). Risk governance: Coping with uncertainty in a complex world. UK, London, Routledge.

Rubinstein, I. S. (2013). Big data: The end of privacy or a new beginning? International Data Privacy, 3(2), 74–86.

Saaty, T. L. (1990). How to make a decision: The analytic hierarchy process. European Journal of Operational Research, 48(1), 9–26.

Şahin, B. D., & Özüdoğru, H. (2019). KOBI'lerde Üretim ve Pazarlama Sorunları: Ostim Örneği 1. Third Sector Social Economic Review, 54(1), 320–333.

Sawka, K. (2000). Are we valuable? Competitive Intelligence Magazine, 3(2), 53–54.

Schick, A., Frolick, M., & Ariyachandra, T. (2011). Competing with BI and Analytics at Monster Worldwide. In Proceedings of the 44 Hawaii International Conference on System Sciences, Hawaii.

Scholz, P., Schieder, C., Kurze, C., Gluchowski, P., & Böhringer, M. (2010). Benefits and Challenges of Business Intelligence Adoption in Small and Medium-Sized Enterprises. European Conference on Information Systems, ECIS 2010, 1–12. https://doi.org/ECIS2010-0252.R1

Scott, R. (1995). Institutions and organizations. Thousand Oaks, GA: Sage.

Seker, S. E. (2016). İş Zekası (Business intelligence). YBS Ansiklopedi, 3(1), 21–25.

Taşgit, Y. E., & Torun, B. (2016). Yöneticilerin İnovasyon Algısı, İnovasyon Sürecini Yönetme Tarzı ve İşletmelerin İnovasyon Performansı Arasındaki İlişkiler: KOBİ'ler Üzerinde Bir Araştırma. Yönetim Bilimleri Dergisi, 14(2), 121–156.

Tekin, E., & Nas, T. İ. (2017). Firma ve Üst Düzey Yönetici Özelliklerinin Türkiye'de Faaliyette Bulunan KOBİ'lerin İhracat Performansı Üzerindeki Etkisi. Ankara Üniversitesi SBF Dergisi, 72(4), 1185–1217.

Thomas Jr, J. H. (2001). Business intelligence–why? eAI Journal, 2001, 47–49.

Turban, E., Sharda, R., Delen, D. (2014). Decision support and business intelligence systems. Pearson Education Limited.

TurkStat. (2018). Survey on Information and Communication Technology (ICT) Usage in Enterprises, 2014-2018.

Ul-Ain, N., Vaia, G., & DeLone, W. (2019, January). Business intelligence system adoption, utilization and success-A systematic literature review. In Proceedings of the 52nd Hawaii International Conference on System Sciences.

Villamarín, J. M., & Diaz Pinzon, B. (2017). Key success factors to business intelligence solution implementation. Journal of Intelligence Studies in Business, 7(1), 48–69.

Visinescu, L. L., Jones, M. C., & Sidorova, A. (2017). Improving decision quality: The role of business intelligence. Journal of Computer Information Systems, 57(1), 58–66.

Wieder, B., & Ossimitz, M. L. (2015). The impact of business intelligence on the quality of decision making–A mediation model. Procedia Computer Science, 64, 1163–1171.

Yeoh, W., & Popovič, A. (2016). Extending the understanding of critical success factors for implementing business intelligence systems. Journal of the Association for Information Science and Technology, 67(1), 134–147.

8 The Development of Loyalty Programs in the Retail Sector

Muniba Rauf
School of Business and Economics,
University of Jyväskylä, Jyväskylä, Finland

Heikki Karjaluoto
School of Business and Economics,
University of Jyväskylä, Jyväskylä, Finland

CONTENTS

8.1 INTRODUCTION

Loyalty programs have significantly evolved since the 1980s, initially considered mere a promotion, Sharp (1997). They are now vital tool for customer relationship management (CRM).

As the economic crisis hits in the 2000s, retailers began refining and redefining their approach to promotions and other marketing tactics by using customers' information and including behavior pattern gathered from loyalty cards data and digital channels (Kang et al., 2015).

Loyalty programs are thus a topic of substantial interest and have become the center of research, particularly regarding membership requirement design, the structure of rewards and points, communication of programs, etc. However, it has been assumed that most loyalty programs are designed from scratch (Breugelmans et al., 2015; Dorotic et al., 2014). In the current market scenario of the retail industry, where most companies already have a loyalty program and are further adding suitable digital channels to their marketing strategies, the core challenge for organizations is revitalizing current loyalty programs and aligning them with their organization's digital activities and assets to gain advantages over their competitors (Liu & Yang, 2009). Additionally, with the increasing trend of implementing loyalty programs in business sectors, various studies have emphasized certain problems regarding their online migration such as catering the needs of customers who do not have mobile internet access (Breugelmans et al., 2015).

Although, many previous studies have examined the relationship between loyalty programs, loyalty cards, and customers' loyalty, more studies are needed because there is still scant evidence about customer's behavior toward loyalty programs and its digitization. Various studies have explored this topic including the rich literature that defines the economies of such loyalty programs (e.g., they are driven by their impact on sales and repurchase behavior); however, most of the available research has either included contradictory evidence and/or has not discussed the impact of loyalty programs on repurchase behavior (Martenson, 2007). Therefore, previous researchers have suggested that much has yet to be done to understand various aspects of customers' behavior toward loyalty programs (Bolton et al., 2000). Table 8.1 presents prior studies on loyalty programs, from which different views and perspectives can be extracted to better understand how loyalty programs' needs have changed over time as well as how the programs have been shaped over the past few years.

Furthermore, a number of researchers (Breugelmans et al., 2015; Purohit & Thakar, 2018) have encouraged the exploration of modern technology usage regarding loyalty programs. Presently, few studies can be found on this topic. In 2016, Berezan et al. studied how digital channels can be used in the hospitality industry including in their loyalty programs. This particular study revealed that the channel that is chosen for sharing information affects customers' perceptions of the information style and quality as well as their loyalty toward the program. The study also suggested that customers find website easy to use and understandable and argued that the importance of social media as a communication channel can never be ignored as it serves as "the fastest catalyst and the best ambassador for instant communication information dissemination" (Berezan et al., 2016, p.111).

Purohit and Thakar (2018) explained that past studies on technology regarding loyalty programs either focused on generic use and recommendations or provided minimal examples of technologies to be used in loyalty programs. Additionally, they observed that the pre and post implementation stages of loyalty programs as well as how the latest technology can be utilized in each stage have not been explored at all (Purohit & Thakar, 2018).

TABLE 8.1
Past Studies on Loyalty Programs (LP)

Year	Authors	Industry	Findings
2016	Steinhoff and Palmatier	Cross-sector	LP can have negative effects on bystander customers, observing other's preferential treatment. LP effectiveness is influenced by reward delivery (rule clarity, reward exclusivity, reward visibility).
2016	Wang et al.	Service industry	LP goal attainment positively impacts post-promotion purchases, whereas goal failure significantly reduces post-purchases.
2014	Dorotic et al.	-	Redemption of LP rewards positively impacts LP members' behavior before and after redeeming a reward.
2012	Kopalle et al.	Service industry	LP design characteristics (frequency of rewards and customer tier component) generate incremental sales without cannibalizing each other.
2009	Liu and Yang	Airline industry	Only high-share firms experienced sales lifts from their loyalty programs. Because high-share firms tend to possess complementary product and customer resources, they are more likely to gain from their loyalty programs than firms with a smaller market share.
2008	Demoulin and Zidda	Grocery industry	Customers satisfied with the rewards of LPs are more loyal to the store and allocate a higher proportion of their budget and patronage frequency to the store than unsatisfied customers.
2008	Bridson et al.	Retailing	LP was a significant predictor of store loyalty, in support of the contention that loyalty programs are capable of engendering loyalty.
2007	Meyer-Waarden	Retailing	LPs have a positive effect on customer lifetime and share of customer expenditures at the store level.
2007	Hennig-Thurau and Paul	Restaurant	LP can lead to counter-productive results by decreasing customer retention.
2007	Liu	Retailing	Positive influence of LP on consumers' purchase frequency and transaction size holds only for light and moderate buyers.
2007	Leenheer et al.	Dutch supermarket industry	Small, positive, yet significant effect of loyalty program membership on share-of-wallet. In terms of profitability, each program generates more additional revenues than additional costs in terms of saving and discount rewards.
2006	Kivetz et al.	Coffee and music on internet	LP induces purchase acceleration through the progress toward a goal.
2006	Gomez et al.	Grocery	LP members are more behavioral and affectively loyal than other participants. Few customers change purchase behavior after joining the program.

(Continued)

TABLE 8.1 *(Continued)*
Past Studies on Loyalty Programs (LP)

Year	Authors	Industry	Findings
2005	Taylor and Neslin	U. S. grocery	LP increases sales through point pressure (short-term) and rewarded behaviors (long-term).
2004	Lewis	Grocery	LPs are successful in increasing repeat-purchase rates.
2003	Reinartz and Kumar	Grocery industry in France	Being a LP member does not modify purchase behavior. Events and promotions associated with LP seem to have clear effects on purchase behavior (e.g., purchase acceleration). The effects of LP are mostly short rather than long term. Thus, they seem to work as promotional tools rather than a means to induce loyalty.
2003	Verhoef	Financial services	LP that provides economic benefit has a positive effect on customer retention and customer share development.
2003	Magi	Retailing	Loyalty cards have mixed effects on consumer behavior (share of purchase and share of visits).
2001	Rajiv	General merchandise	LP membership is associated with the longer duration of customer–firm relationships.
2001	Meyer-Waarden and Benavent	U.S. grocery industry	LP is operationalized as a shocker program (e.g. turkey bucks), not a traditional long-term card program, so it can better be described as a long promotion. There is significant increase in spending (market basket). LPs seem to affect "cherry-pickers" most. Program is profitable.
2000	Bolton et al.	General retail	LP has hardly any effect on repeat purchase patterns (behavioral loyalty).
2000	Deighton and Shoemaker	Credit cards	LP members are more likely to overlook negative experiences with the focal company. LP members have higher usage levels and higher retention.
2000	Crie et al.	Hospitality	20% of member stays are because of LP. Strategy of using LP as a value alignment tool is successful. LP is profitable.

Therefore, this chapter aims to focus on the following: (1) typical tasks that are performed at each stage of the loyalty program lifecycle, (2) recommendations based on research that can enhance the effectiveness of the loyalty program, and (3) the use of relevant technology to achieve the needed tasks.

8.2 LITERATURE REVIEW

The below section of this chapter discusses the relevant past studies on the topic of loyalty programs, the design of loyalty programs and their findings.

8.2.1 Loyalty Programs

A loyalty program is suggested to induce positive feeling that encourages members to make repeated purchases. Kivetz and Simonson's (2002) loyalty programs generate the feeling of pride about either the achievement (of rewards/points) or winning something without paying extra among customers.

Customer loyalty is the core objective of a company and its relationship marketing (Palmatier et al., 2006). Thus, companies in all industries opt for loyalty programs to build and enhance customer relationships (Kivetz & Simonson, 2002; Nunes & Dreze, 2006).

Loyalty programs are introduced with the purpose of encouraging customers to visit and make purchases (Demoulin & Zidda, 2008). Loyalty programs consist of an integrated system of marketing communication and actions with the purpose of increasing loyalty, repeat purchase behaviors, switching costs through providing economical functions, and informational and sociological rewards (Meyer-Waarden, et al., 2008).

Therefore, a customer's brand loyalty is an important factor for growing business, making it important to create marketing strategies that will appeal to each customer on an individual level.

8.2.2 Traditional Loyalty Programs

The simplest and one of the most traditional systems, card loyalty systems, is the punch card that is usually given to customers for free. After every purchase of a specific item (such as coffee), a hole is punched. When a fixed number of purchases are completed, the customer receives either a free gift or a discount. These types of card loyalty programs are easy to implement and therefore most commonly used by small vendors or retailers.

Currently, the concept of traditional loyalty programs, where consumers are only rewarded for their in-store purchases, has been diluted and thus given way to more sophisticated and modern loyalty card programs, which have become a hot topic among marketers and retailers. Notably, the emergence of smart phones and other digital gadgets as well as social media are making traditional loyalty programs obsolete; they no longer meet the demands of modern customers who often expect synergy across all channels (Driscoll, 2013; Verhoef et al., 2015).

There is increased interest in retaining old customers rather than acquiring new ones and a rise in competition in the retail sector. Therefore, retailers are offering various forms of loyalty programs to boost sales and increase brand recognition by striving for customer repurchases over time. Even though there are many loyalty programs in the retail sector, few studies have focused on measuring the influence such programs on customer loyalty and retention regarding technology and their digitization (Table 8.1).

8.2.3 Loyalty Programs and its Technology use

It has been noted that while researchers provide strong and continuous recommendations for the use of digital technology in loyalty programs (Breugelmans et al., 2015; McCall et al., 2010), studies that address the use of appropriate technology in loyalty programs are quite limited and focus on few technologies (Liljander et al. 2007; Son et al., 2016; Wang et al., 2016).

As per Bijmolt et al. (2011), Liu and Yang (2009), and McCall et al. (2010), the latest advancements and trends in information technology and communication, marketing analytics, and consumer interface platforms such as mobile devices, have offered recent developments in loyalty program practices while providing new opportunities as well as challenges because, although it attracts the early adapters to the program and technology, it is only beneficial to those customers who use the mobile internet and 4G (see Figure 8.1). Son et al. (2016) suggested that such mobile applications and engagement can have a positive effect on cash expenditure and point redemption.

8.3 THE LOYALTY PROGRAM LIFECYCLE: DESIGN, IMPLEMENTATION AND ASSESSMENT

8.3.1 The Design Stage

If a loyalty program's design lacks appropriate planning and attention, the plan may be ineffective. On the contrary, a well-planned and well-designed loyalty program attracts customers and dissuades them from joining competitors' programs (Meyer-Waarden, 2008). As discussed by McCall et al. (2010), there is no specific rule for designing a loyalty program, but there are crucial points that demands attention, including overall structure, point structure, membership requirements, reward structure, and program communication. An important decision is whether a program should be made open to all or offered only to a specific group of customers. Both have pros and cons, depending on the need of the organization. For instance, an open-to-all loyalty program would create program awareness and bring benefits to its customers (Breugelmans et al., 2015). By contrasts, an invitation-only loyalty program would make efficient use of the available budget by excluding customer groups that offer low profitability.

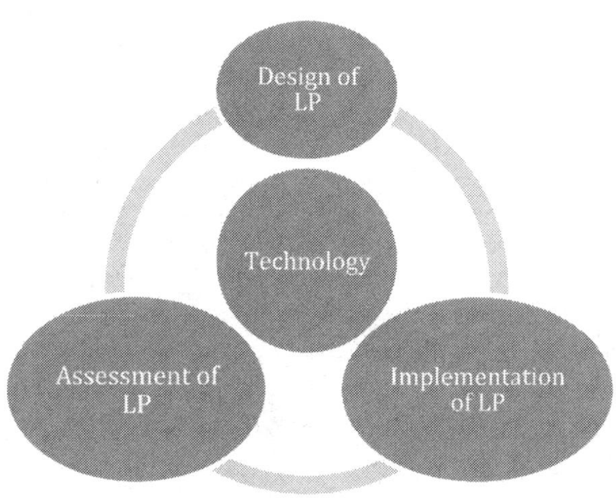

FIGURE 8.1 Technology and lifecycle of a loyalty program.

Similarly, decisions in terms of the reward structure are also made during the design phase. These include the number of tiers, points' expiry rules, points issuance ratio, tier benefits, tier positioning, and tier transition rules. Depending on the need and nature of the program, an organization may choose either reward option (customer tier or frequency reward), or it may opt for a combination of both (Kopalle et al., 2012). Doroctic et al. (2014) explained that companies should refrain from point expiry and/or binding thresholds to positively influence purchase behavior. Consumers' spending habits over a certain period have usually been associated with tier upgrades. Nevertheless, companies may upgrade the status of a specific member to a higher tier even before he/she attains it (through purchases). This would result in feeling of skepticism and gratitude. In such cases, companies should only select customers who are already close to achieving the higher tier/status and the customer should be given a choice beforehand (Eggert et al., 2014).

The constructs of loyalty programs' stages such as design, implementation and assessment are prominent in prior literature as shown in Table 8.2.

Furthermore, it is important to determine appropriate metrics, such as customer retention, attitudinal loyalty, and repeat purchase when designing a loyalty program. It would further help to evaluate program performance as well as determine the return on investment (ROI) (Balakrishnan, 2011).

TABLE 8.2
Stages of the Loyalty Program Lifecycle and Their Corresponding References

Stage	Reference	Concept
Design	Bijmolt et al. (2010)	"Loyalty programs: Generalizations on their adoption, effectiveness, and design" (p. 197). "In general, an LP design comprises the program's structure, rewards, and a number of program partners." (p. 230).
	Kreis and Mafael (2014)	"The influence of customer loyalty program design on the relationship between customer motives and value perception" (p. 590).
Implementation	Kongarchapatara and Shannon (2012)	Investigating the effectiveness of a loyalty program through the relationships of program design, implementation, and customer loyalty" (p. 1).
	Shugan (2005)	"The design and implementation of loyalty programs is both an important and growing area of research." (p. 191).
Assessment	Breugelmans et al. (2015)	"Assessment of LP performance" (p. 132).
	Kang et al. (2015)	Beyond traditional behavioral measures, CCID and company latent financial risk offer alternative assessments of (LP) effectiveness." (p. 468).

From the literature above, it can be seen that the crucial decisions of loyalty program design are based on the information and/or data that can be extracted from a number of sources distributed across channels such as the company's enterprise systems, partners, industry, and the market. Apart from traditional systems and sources, the required data can also be obtained from modern sources, such as social media and web logs, and it may be used effectively in the decision- making process to provide relevant program and offers to customers (Purohit & Thakar, 2019). Thus, the critical role of technology in the design stage of a loyalty program involves retrieving, transforming, storing, and processing large amounts of data that comes in different forms and from different systems. Data processing steps include extraction from the main source, validation, transformation of the data into the appropriate format for storage, and loading it into storage for future use. Therefore, extraction, transformation, and loading tools (e.g. Informatica) may be beneficial for automating data handling (Purohit & Thakar, 2019).

Purohit and Thakar (2019) also suggested that after preparation of the required data, it is possible to use simulation programs to predict the potential outcomes of a certain choice in the design using a what-if analysis. This can help retailers and companies save both time as well as cost. Similarly, the dynamic demand for computation resources during the design phase may prove the use of cloud-based services economical.

The overall outcome of the design stage can be seen in Figure 8.2.

8.3.2 The Implementation Stage

The implementation stage of any loyalty program is at least equal to, if not more than the design phase, in creating an effective loyalty program (Xie & Chen, 2013). The implementation process should coexist with and match the relevant parameters of the design process. The key activities and tasks as well as the role of technology in the implementation process are discussed below.

8.3.2.1 Communication

A well-known marketing statement suggests that it is better to retain an existing customer than to acquire a new one because it results in repetitive interactions. The retailer then becomes more familiar with such customers and can offer tailored promotions and products. For instance, cosmetic retailer Sephora creates a beauty profile for each member of its loyalty program. The specific profile then generates personalized offers and recommendations based on skin and hair types (Colloquy, 2015) to

FIGURE 8.2 Technology and design stages of loyalty program.

give members maximum benefit of their products. This further allows Sephora to have a competitive edge over other beauty retailers.

Another retail champion with a personalized loyalty program is Tesco. Their Clubcard, which was launched in 1995, was a huge success pushing Tesco ahead of Sainsbury to become the market leader in the UK grocery sector (Marketing Week, 1995). The company that created the loyalty program (Dunnhumby) used purchase information to create customer profiles and they later built a model that could accurately predict future consumptions and needs of customers. Additionally, the Coupon At Till (CAT) system brings Tesco customers back to stores via personalized offers that are appealing and relevant (Humby et al., 2004). Over three month-period in 2005, Tesco sent six million personalized combinations of coupons to its customers (Tesco, 2005).

For the casual personalized offer strategy to work both efficiently and effectively, attracting customers to a store is not enough. The other significant step is to ensure that customers not only purchase items for which they receive coupons/offers, but also other items in the store (Nastasoiu & Vandenbosch, 2019). Therefore, with effective personalized, the following two strategic decisions are crucial: (1) whom to target, and (2) what type of offers and promotions to send out. The first decision is the most important because the wrong decision can lead to adverse situations, such as offering incentives and promotions to those who would purchase the offered items anyway. The second decision requires an accurate match between the customer and the offers/promotions. Per Nair et al., (2017), intelligent matching can increase revenue significantly, without necessarily increasing cost.

Wiebenga and Fennis (2014) revealed that customers' behavior can be influenced by making changes in the way the progress of the program is communicated.

8.3.2.2 Communication Style

Regarding a loyalty program's success, program communication plays a significant role. The communication style as well as the quality of the information offered affects customers' commitment to the program (Ball et al., 2004; Sharma & Patterson, 1999). Additionally the channel through which the information is being communicated plays a vital role in how program members perceive the style of communication and the information's quality (Berezan et al., 2016). For example, websites and social media are known to improve a store/brand's image and to significantly affect the behavior of customer toward the loyalty program (Liljander et al., 2007; Son et al., 2016). Therefore, it is crucial that the retail stores provide relevant, consistent and accurate information and message across all touchpoints and channels to enhance customers' perception of their loyalty programs (Liljander et al., 2007).

Additionally, the style of communication and its content affect all key areas of relationships between a firm and its customers, including loyalty (Ball et al., 2004). During the early stages, communication helps build brand awareness, develop brand preferences, influence current customers, and encourage potential customers to make purchases (Ndubisi & Chan, 2005). During the later stages, communication provides the opportunity to maintain regular contact with customers to provide real time and accurate information and updates on products and services, as well as to proactively address potential problems and find solutions. In terms of loyalty programs, communication can be either firm or customer-created. Studies have found that, by supporting both of these communications types, social media and other touchpoint platforms can influence intention to spread positive electronic word of mouth (eWoM) and thus create and enhance loyalty (Zhang & Benyoucef, 2016).

8.3.3 Firm-Created Communication

The retail and service industry's abilities to meet or even exceed customers' needs have been greatly improved by ever-changing communication media, expectations of accurate and personalized information, and expanded customer touchpoints (Ray et al., 2005). Companies are striving to use all touchpoints through various methods, including personalization, such as direct emails and mail, personalized letters, interaction via websites, and machine-generated interaction as well as personal interaction between the company and its customers via the pre-selling, consumption, and post-selling stages (Ball et al., 2004; Zahay et al., 2014).

For this reason, a successful loyalty program must create an interpersonal connection between the company and its customers through solid communication (Shoemaker & Lewis, 1999). Numerous studies have focused on the importance of personalized communication in order to gain customers' loyalty (Allen & Wilburn, 2002; Lemon et al., 2001). If an organization manages data properly, and couples that with technology, such as Internet and other platforms, it can lead highly personal communication (Zahay et al., 2012).

8.3.4 Customer-Created Communication

Customer-created communication includes both electronic as well as traditional WoM. It involves interactions between customers on various social media platforms which can enhance a customer's knowledge base and can maximize his/her benefits from the loyalty program. This can also enhance the overall experience and his/her perceived value of the program. Gruen et al. (2006) found that WoM has been perceived by customers as a trustworthy source of information. Online travel forums like yelp.com, tripadvisor.com, and flyertalk.com offer their members service ratings, discussion forums, and reviews that allow the customers to have interactive conversations. Berezan et al. (2015) stated that an important aspect of effective communication is its style including whether it is personalized and interactive, per the customer's perception.

8.3.4.1 Customer Support

While companies move ahead with good intentions and precautions, negative incidence can occur either when customers are using services or at the time of purchase. Stauss et al. (2005) argued that these incidents can cause frustration as well as negative reactions among customers regarding the loyalty program and the brand overall. However, these incidents can be successfully addressed if they are integrated well with customer support systems that can be used to assign the appropriate priority to the problem based on the member's tier and profile. For example, compensation points can be offered to reassure customers' loyalty.

8.3.4.2 Privacy Matters

Around the world, organizations are now using data that was obtained from the profile of the members to provide personalized offers. This can cause mistrust amongst customers, they may not feel confident and comfortable enough sharing personal information as they fear its misuse (Ashley et al., 2011). Furthermore, as customers have become more aware of privacy concerns and security issues, the acceptance of loyalty programs has been negatively affected (Blanco-Justicia & Domingo-Ferrer, 2016).

One solution to this problem is that companies can employ information security control and related policies. This involves using improved authentication procedures such as Captcha codes on every digital channel that is used by customers. Companies can also opt for a multitier privacy control mechanism (Blanco-Justicia & Domingo-Ferrer, 2016; Enzmann & Schneider, 2005), which would lead to building confidence in the program, among its members.

8.3.4.3 Location Based Services

Location-based services are proving to be beneficial for retailers by allowing them to provide accurate and relevant offers to their respective customers (Brynjolfsson et al., 2013). Additionally, as suggested previously, such services may be well integrated into retailers' digital loyalty programs in order to provide relevant and personalized offers through considering profile specifics, such as gender, age, buying behavior, and buying history. Retailers can also use these location-based services to detect and identify potential loyal customers who merely visit the store without making a purchase. The retailer can then initiate customized offers to such members in order to retain them and gain their loyalty.

8.3.4.4 Automation and Efficiency

Staff at the member support centers of any reward program tends to perform a number of operational tasks on a regular basis, including progressing redemption requests, managing the fulfillment of promotional merchandise, monitoring member activity statements, and/or processing the complaints and queries. Most of these tasks involve predefined procedures. On this point, Brynjolfsson and Mitchell (2017) elaborated that some tasks can be automated by evaluation of the applied criteria. In such circumstances, artificial intelligence (AI) can be used to observe and learn human decisions. By implementing automation to the routine tasks of their loyalty programs, organizations can increase efficiency, cut operational costs, and utilize these freed up resources for other strategic tasks.

8.3.5 The Performance Assessment Stage

There are various performance measures for loyalty programs, which may include customer traffic, enrolment, frequency of purchase, share of wallet, WoM, reduced price sensitivity, etc. (Breugelmans et al., 2015; McCall et al., 2010). However, the pre and essential steps to measure the ROI are to define suitable metrics that are aligned with the objectives of the firm or which the loyalty program was adopted (Balakrishnan, 2011).

Some studies (Breugelmans et al., 2015; Pauwels et al., 2009) have suggested a dashboard approach to assess a loyalty program. This approach is possible if the firm: (1) defines and identifies the appropriate metrics during the design phase, (2) collects accurate elements of the data for the metrics at implementation phase, and (3) extracts, transforms, and loads the data into a data warehouse for the purpose of analysis.

A study by Frisou and Yildiz (2011) supported the use of technology in loyalty programs by emphasizing that the effectiveness of such programs is associated with the learning of the consumer; consumers spend more on purchases if and when they learn about and are fully aware of the accumulation and redemption of points. Therefore, if the "learning" is used as a metric during the design stage to improve program's effectiveness, there should also be some mechanism in place to measure it. This can be done in a variety of known ways like via questionnaires, quizzes, and/or surveys. As an alternative, indirect methods can also be used to measure it, including measuring when a customer performs a particular activity over a prolonged period. Information technology can be used for both of these approaches and measures. The task of each stage with their respective outcomes can be seen in Table 8.3.

To summarize, business intelligence (BI) software and data warehouses can be used effectively to measure the performance of any loyalty program. However, the artificial intelligence (AI) can be used beyond the assessment of standard procedures to efficiently reveal patterns of customer behavior, which may not be otherwise obvious; by the application of a machine learning algorithm on data of the members. Trends can then be analyzed in order to identify loyalty program members' characteristics which can further help companies to predict loyal versus non loyal customers. With such insight, the firms can take proactive actions in order to retain customers through loyalty programs.

TABLE 8.3
Tasks and Outcomes of Stages

Stages	Tasks	Action	Outcome
Design	• Defining LP objectives • Development of budget • Determining LP eligibility • Selecting LP reward		Blueprint of LP with rules and parameters
Implementation	• Building a relevant organization • Developing and maintaining a LP database • Managing internal data warehourse and data mining	• Effective communication • Personalization • Privacy rules • Automation and efficiency	Effective transition of design into action using technology
Assessment	• Evaluating the LP • Taking concrete actions		Effectiveness metrics, gaps and concrete actions

8.4 DISCUSSION

Although loyalty programs in retail industry are still considered the most important tool for CRM; there is no denying that they are reaching maturity. However, recent advancements in technology can influence all stakeholders including customers, companies as well as partner organizations (if any). Thus, modern technology and information can play vital roles in loyalty programs and their elements that can enable firm to continue providing value and gain greater ROI.

This chapter explains the benefits of using technology and information in the loyalty program lifecycle in order to enhance the overall effectiveness of loyalty programs and thus provides the following contributions:

- Demonstrates the tasks that are being performed at each phase of loyalty program lifecycle.
- Provides an extensive list of technology for the program lifecycle, including specific technologies that can be useful for each stage.
- Demonstrates how the technology can be implemented in each phase of program's lifecycle to achieve goals and increase effectiveness.
- Explains how the use of technology in loyalty programs can effectively retain the customers and enhance their loyalty toward the program as well as the brand.

Technology in loyalty programs is still a vast field that requires additional exploration and discovery. While it is beneficial to encourage customers to consider the loyalty program at each visit to the store; complete integration and personalization are next steps. Technology will soon allow tracking of all loyalty cards in one wallet, intuitive points' redemption without the need to even select relevant loyalty card, and the authority to exchange points across multiple loyalty programs. In this way, organizations can gain better insight into customer behavior and can use the obtained information to predict behaviors as well as outcomes. However, to achieve this stage, an open business model and partnership with mobile wallet providers are key to success.

REFERENCES

Allen, D. R. & Wilburn, M. (2002). Linking Customer and Employee Satisfaction to the Bottom Line. American Society for Quality, Milwaukee, WI.

Ashley, C., Noble, S., Donthu, N., & Lemon, K. (2011). Why customers won't relate: Obstacles to relationship marketing engagement. Journal of Business Research, 64(7), 749–756.

Balakrishnan, M. S. (2011). Gain the most from your marketing spend. Business Strategy Series, 12(5), 219–225.

Ball, D., Simões Coelho, P., & Machás, A. (2004). The role of communication and trust in explaining customer loyalty: An extension to the ECSI model. European Journal of Marketing, 38(9/10), 1272–1293.

Berezan, O., Yoo, M., & Christodoulidou, N. (2016). The impact of communication channels on communication style and information quality for hotel loyalty programs. Journal of Hospitality and Tourism Technology, 7(1), 100–116.

Bijmolt et al., (2011). Loyalty programs: Generalizations on their adoption, effectiveness and design. Found Trends Marketing, 5(4), 197–258.

Bolton, R. N., Kannan, P. K., & Bramlett, M. (2000). Implications of loyalty program membership and service experiences for customer retention and value. Journal of Academy of Marketing Science, 28, 95–108.

Blanco-Justicia, A., & Domingo-Ferrer, J. (2016). Privacy-aware loyalty programs. Computer Communication, 82, 83–94.

Breugelmans, E. et al., (2015). Advancing research on loyalty programs: A future research agenda, Marketing Letters, 26(2), 127–139.

Bridson et al., (2008). Assessing the relationship between loyalty program attributes, store satisfaction and store loyalty. Journal of Retailing and Consumer Services, 15(5), 364–374.

Brynjolfsson, E., Hu, Y. J., & Rahman, M. S. (2013) Competing in the age of omni channel retailing. MIT Sloan Management Review, 54(4), 23.

Brynjolfsson, E., & Mitchell, T. (2017). What can machine learning do? Workforce implications. Science, 358(6370), 1530–1534.

COLLOQUY, (2015). The 2015 COLLOQUY Loyalty Census: big numbers and big hurdles, Available at: www.colloquy.com/latest-news/2015-colloquy-loyalty-census/

Crié et al., (2000). Analysis of the efficiency of loyalty programs. Working Paper, University of Pau, France.

Daikoku, G. (2012) Retail loyalty management: More than points and reward. Stamford CT, Gartner.

Deighton, J., & Shoemaker, S. (2000). Hilton Honors worldwide: Loyalty wars. Harvard Business. https://www.hbs.edu/faculty/Pages/item.aspx?num=27553.

Dorotic et al., (2014) Reward redemption effects in a loyalty program when customers choose how much and when to redeem. International Journal of Research in Marketing, 31, 339–355.

Demoulin, N., & Zidda, P. (2008). On the impact of loyalty cards on store loyalty: Does the customers' satisfaction with the reward scheme matter?. Journal of Retailing & Consumer Services. https://doi.org/10.1016/j.jretconser.2007.10.001

Driscoll, M. (2013). Bye-bye, silo retailing, hello, omni-channel. Value Retail News, 31(4), 10–14.

Eggert, A., Garnefeld, I., & Steinhoff, L. (2014). How profound is the allure of endowed status in hierarchical loyalty programs? In A. Geyer-Schulz, & L. Meyer-Waarden (Eds). Customer and service systems (p. 31). Karlsruhe, KIT Scientific Publishing.

Enzmann, M., & Schneider, M. (2005). Improving customer retention in e-commerce through a secure and privacy-enhanced loyalty system. Information Systems Frontier, 7(4), 359–370.

Frisou, J., Yildiz, H. (2011). Consumer learning as a determinant of a multi-partner loyalty program's effectiveness: A behaviorist and long-term perspective. Journal of Retailing and Consumer Services, 18(1), 81–91.

Gómez et al., (2006). The role of loyalty programs in behavioral and affective loyalty. Journal of Consumer Marketing, 23(7), 387–396.

Gruen, T. W., Osmonbekov, T., & Czaplewski, A. J. (2006). e-WOM: The impact of customer- to-customer online know-how exchange on customer value and loyalty, Journal of Business Research, 59(4), 449–456.

Hennig-Thurau, T., & Paul, M. (2007). Can economic bonus programs jeopardize service relationships? Service Business, 1(2), 159–175.

Humby et al., (2004). Scoring points: How Tesco is winning customer loyalty. London, UK: Kogan Page.

Kang, J., Brashear-Alejandro, T., & Groza, M. (2015). Customer–company identification and the effectiveness of loyalty programs. Journal of Business Research, 68(2), 464–471.

Kivetz, R., & Simonson, I. (2002). Earning the right to indulge: Effort as a determinant of customer preferences toward frequency program rewards. Journal of Marketing Research, 39, 155–170.

Kivetz et al., (2006). The goal-gradient hypothesis resurrected: Purchase acceleration, illusionary goal progress, and customer retention. Journal of Marketing Research, 43(1), 39–58.

Kopalle, P., Sun, Y., Neslin, S., Sun, B., & Swaminathan, V. (2012). The joint sales impact of frequency reward and customer tier components of loyalty programs. Marking Science, 31(2), 216–235.

Kongarchapatara, B., & Shannon, R. (2012). Investigating the effectiveness of a loyalty program through the relationships of program design, implementation, and customer loyalty. ANZMAC2012. Adelaide

Kreis, H., & Mafael, A. (2014). The influence of customer loyalty program design on the relationship between customer motives and value perception. Journal of Retail Consumer Service, 21(4), 590–600.

Leenheer et al. (2007). Do loyalty programs really enhance behavioral loyalty? An empirical analysis accounting for self-selecting members. International Journal of Research in Marketing, 24(1), 31–47.

Lemon, K. N., Rust, R. T. & Zeithaml, V. A. (2001). What drives customer equity?, Marketing Management, 10(1), 20–25.

Lewis, M. (2004). The influence of loyalty programs and short-term promotions on customer retention. Journal of Marketing Research, 41(3), 281–292.

Liljander, V., Polsa, P., & Forsberg, K. (2007). Do mobile CRM services appeal to loyalty program customers? International Journal of E-Business Research (IJEBR), 3(2), 24–40.

Liu, Y. (2007). The long-term impact of loyalty programs on consumer purchase behavior and loyalty. Journal of Marketing, 71(4), 19–35.

Liu, Y., & Yang, R. (2009). Competing loyalty programs: Impact of market saturation, market share, and category expandability. Journal of Marketing, 73, 93–108.

Mägi, A. W. (2003). Share of wallet in retailing: The effects of customer satisfaction, loyalty cards and shopper characteristics. Journal of Retailing, 79(2), 97–106.

Marketing Week. (1995, April 14). Tesco's sales overtake Sainsbury's for the first time. Available at: https://www.marketingweek.com/tescos-sales-overtake-sainsburys-for-first-time/

Martenson, R. (2007). Corporate brand image, satisfaction and store loyalty: A study of the store as a brand, store brands and manufacturer brands. International Journal of Retail & Distribution Management, 35(7), 544–555.

McCall, M., & Voorhees, C. (2010). The drivers of loyalty program success: An organizing framework and research agenda (Electronic version). Cornell Hospitality Quarterly, 51(1), 35–52.

Meyer-Waarden, L., & Benavent, C. (2001). Loyalty programs: Strategies and practice. Working Paper, University of Pau, France.

Meyer-Waarden, L. (2007). The effect of loyalty programs on customer lifetime duration and share of wallet. Journal of Retailing, 83(2), 223–236.

Meyer-Waarden, L. (2008). The influence of loyalty program membership on customer purchase behaviour. European Journal of Marketing, 42(1/2), 87–114.

Nair, S., Misra, S., Hornbuckle, W. J., IV, Mishra, R., & Acharya, A. (2017). Big data and marketing analytics in gaming: Combining empirical models and field experimentation. Marketing Science, 36(5), 699–725.

Nastasoiu & Vandenbosch (2019). Competing with loyalty: How to design successful customer loyalty reward programs. Business Horizons (2019), 62, 207–214.

Ndubisi, N., & Chan, K. (2005). Factorial and discriminant analyses of the underpinnings of relationship marketing and customer satisfaction. International Journal of Bank Marketing, 23(7), 542–557.

Nunes, J. C., & Dreze, X. (2006). Your loyalty program is betraying you. Harvard Business Review, 84(4), 124–131.

Palmatier, R. W., Dant, R. P., Grewal, D., & Evans, K. R. (2006). Factors influencing the effectiveness of relationship marketing: A meta-analysis. Journal of Marketing, 70(4), 136–153.

Pauwels, K., Ambler, T., Clark, B., LaPointe, P., Reibstein, D., Skiera, B. et al., (2009). Dashboards as a service. Why, what, how, and what research is needed? Journal of Service Research, 12(2), 175–189.

Purohit, A., & Thakar, U. (2018). Use of ICT in customer loyalty programs: A state of the art review. International Journal Research Applied Science Eng Technology (IJRASET), 6(4), 1010–1020.

Purohit, A., & Thakar, U. (2019). Role of information and communication technology in improving loyalty program effectiveness: A comprehensive approach and future research agenda. Journal of Information Technology & Tourism, 21(2), 259–280.

Rajiv, L. (2001). Retail loyalty programs: Do they work? Marketing Science Conference, Wiesbaden, Germany.

Ray, G., Muhanna, W. A. & Barney, J. B. (2005). Information technology and the performance of the customer service process: A resource-based analysis. MIS Quarterly, 29(4), 625–652.

Reinartz, W., & Kumar, V. (2003). The impact of customer relationship characteristics on profitable lifetime duration. Journal of Marketing, 67(1), 77–99.

Sharp, B. (1997). Loyalty programs and their impact on repeat purchase loyalty patterns. International Journal of Research Marking, 14(5), 473–486.

Sharma, N., Patterson, P. (1999). The impact of communication effectiveness and service quality on relationship commitment in consumer, professional services. Journal of Service Marketing, 13(2), 151–170.

Shoemaker, S., & Lewis, R. (1999). Customer loyalty: the future of hospitality marketing. International Journal of Hospitality Management, 18(4), 345–370.

Shugan, S. M. (2005). Do rewards really create loyalty? Marketing Science, 24, 185–193.

Son, Y., Oh, W., Han, S., & Park, S. (2016). The adoption and use of mobile application-based reward systems: Implications for offline purchase and mobile commerce. In: ICIS 2016 Proceedings.

Stauss, B., Schmidt, M., & Schoeler, A. (2005). Customer frustration in loyalty programs. International Journal of Service Industry Management, 16(3), 229–252.

Steinhoff, L., & Palmatier, R. W. (2016). Understanding loyalty program effectiveness: Managing target and bystander effects. Journal of the Academy of Marketing Science, 44(1), 88–107.

Taylor, G. A., & Neslin, S. A. (2005). The current and future sales impact of a retail frequency reward program. Journal of Retailing, 81(4), 293–305.

Tesco, (2005, August 4). This Sceptered Aisle. The Economist.

Verhoef, P. C. (2003). Understanding the effect of customer relationship management efforts on customer retention and customer share development. Journal of Marketing, 67(4), 30–45.

Verhoef, P. C., Kannan, P. K., & Inman, J. J. (2015). From Multi-Channel Retailing to Omni-Channel Retailing Introduction to the Special Issue on Multi-Channel Retailing. Journal of Retailing and Consumer Services, 91(2), 174–181.

Wang, D., Xiang, Z., Law, R., & Ki, T. P. (2016). Assessing hotel-related smartphone apps using online reviews. Journal of Hospitality Marketing & Management, 25(3), 291–313.

Wang et al., (2016). Enduring effects of goal achievement and failure within customer loyalty programs: A large-scale field experiment. Marketing Science, 35(4), 565–575.

Wiebenga, J. H., & Fennis, B. M. (2014). The road traveled, the road ahead, or simply on the road? When progress framing affects motivation in goal pursuit. Journal of Consumer Psychology, 24(1), 49–62.

Xie, K. L., & Chen, C. C. (2013). Progress in loyalty program research: Facts, debates, and future research. Journal of Hospitality Marking & Management, 22(5), 463–469.

Zahay, D., Peltier, J. & Krishen, A. S. (2012). Building the foundation for customer data quality in CRM systems for financial services firms. Journal of Database Marketing & Customer Strategy Management, 19(1), 5–16.

Zahay, D., Peltier, J., Krishen, A. S. & Schultz, D. E. (2014). Organizational processes for B2B services IMC data quality. Journal of Business & Industrial Marketing, 29(1), 63–74.

Zhang, K. Z. K. & Benyoucef, M. (2016). Consumer behavior in social commerce: A literature review. Decision Support Systems, 86(1), 95–108.

9 Business Intelligence, Big Data and Data Governance

Hélder Quintela
Hight School of Technology
Polytechnic Institute of Cávado and Ave, Barcelos, Portugal

Davide Carneiro
School of Management and Technology,
Polytechnic Institute of Porto, Porto, Portugal

Luís Ferreira
Hight School of Technology, Researcher – 2Ai –
Applied Artificial Intelligence Laboratory,
Polytechnic Institute of Cávado and Ave, Portugal

CONTENTS

9.1 INTRODUCTION

The information always represented a critical factor for knowledge acquisition and advantage on any scenario of dispute of success and competition. The timely transformation of raw data to rich data, and information to rich knowledge, requires strategies and programs that competitive organizations cannot ignore.

There is a considerable investment on such strategies for modeling or implementation. Business Intelligence (BI), Big Data, and Data Analysis are emergent initiatives. However, these investments are not always efficiently planned, or its implementation results measured. Most of the times, the companies "follow" existing business models and other companies' strategies. The business impact is consequences that, most of the times, were not predicted or anticipated.

This research focusses the relevance on having data-centric strategies and emphasizes the transformative requirements that companies need to apply toward an effective intelligence on decision support. Data Governance programs and data maturity assessment are key roles on such transformative business scenario.

If in the recent years the main concern of the organizations were related with adoption of tools and platforms to analyze data properly and efficiently, currently the challenges related with BI, Big Data, Data Science are not only related with technology but also with data governance, data management, data quality, data maturity, and Return on Investment.

Considering this framework, this chapter is organized as follows: first, in order to better position the reader, an overview of BI, Big Data, and Data Science topics is made; their relation, evolution, and applications are placed; and inherent challenges for its quality and value, emphasized; second, follows the exposition and assessment of data maturity and its relevance for BI improvement; and the last section presents the principles about Data Governance.

9.2 FROM BUSINESS INTELLIGENCE TO BIG DATA AND DATA SCIENCE

The corporate world is nowadays an extremely ambitious environment, in which organizations and the people that constitute them maintain a continuous competitive edge to deliver the best they can, as fast as they can. BI initiatives emerged in this milieu as a way to leverage on organizational data to provide valuable organizational and strategic decision support for stakeholders.

The value proposition of BI is thus to transform data into usable knowledge [1]. This transformation process is generally organized in three main sequential phases, each one associated to a major technology: Data Warehouses, Online Analytical Processing (OLAP), and Data Mining.

The first step in this process is to collect, organize, and store the actual raw data that may be internal to the organization as well as external (e.g., competitors). Data is generally stored in a Data Warehouse, which is an integrated data repository that allows not only to store large volumes of data, but also facilitates its use by nonspecialist users. One of its key aspects is to store data along with its different dimensions, which makes operations very intuitive when compared to a traditional database as it allows to visualize and use data according to different perspectives. At this level, businesses can resort to reporting and analysis tools that allow stakeholders to assess and visualize stored data, which represents a past view of the organization, that is, they can answer the question "how did we get here?"

The next step is to analyze the data under the light of the specific domain of the organization, generating information. The main technology in this step is OLAP tools that are supported by the Data Warehouse. OLAP operations can be very diverse and include slicing (selecting data from one dimension/perspective), drilling (navigating up or down in the hierarchical dimensions of the data), and pivoting or rotating (visualize data from a new perspective). In this step, technology mainly enables the generation and visualization of information about the organization in real-time. This allows stakeholders to track business performance, answering the question "what is happening right now?" and taking real-time decision that impact the near future of the organization. These decisions can be supported by dashboards, KPI (Key Performance Indicator) scorecards, and other similar visual tools.

Finally, in the third step, information is transformed into knowledge when the BI users consume it to take decisions. Knowledge can be generated by technologies such as Data Mining, machine learning, or Predictive Analytics, which allow stakeholders to create domain models or to identify patterns or rules. This high-level knowledge becomes an essential element for decision makers to plan ahead and push the organization toward more favorable internal and external conditions, as it allows stakeholders to answer the question "where will we be in the future?"

As we move in this information value chain from beginning to end, we find technologies that enable increased business value but are also inherently more complex to implement and maintain (Figure 9.1).

While the value of BI for the organization may reside in different parts of the information value chain (according to the organization's mission and goals at any given moment), it is generally the third level that provides the highest value as Predictive Analytics technologies allow the organization to plan ahead and prepare for the future.

Nonetheless, a BI initiative, in itself, is not the leverage that will distinguish one organization from the other since its value is highly dependent on the people that interpret the information and take decisions. It rather materializes the potential for the organization to become better – faster, more intelligent, more aware, more creative in solving problems – provided that the stakeholders make an efficient use of the information.

That is, despite all its merits, BI remains just another tool that may eventually support all the levels of the organization. Its success may be tied to many other

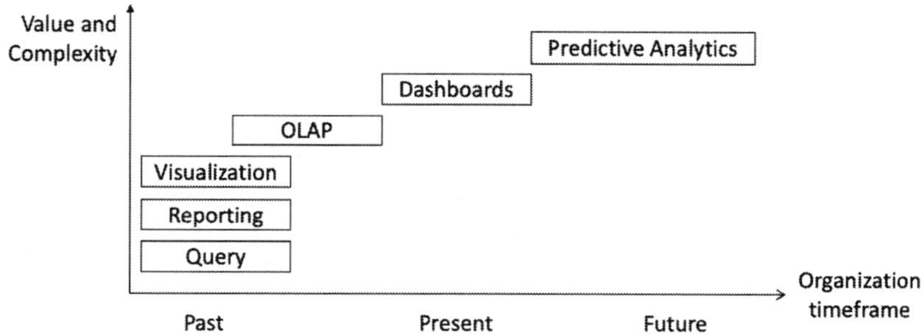

FIGURE 9.1 BI technologies in relation to the organization timeframe and their value and complexity.

factors including the existence of adequate human resources management, a proper governance framework, collaborative working environments oriented toward group-based decision-support, and an overall focus on communication, problem-solving, and risk and failure tolerance. Otherwise, the implementation of a BI initiative may become just a technological and financial burden for the organization.

Moreover, a BI solution cannot be implemented in a traditional fashion, as would happen with the implementation of other systems [2]. BI is rather exploratory: it entails a discovery process that takes place continuously, often as the solution is implemented. For this reason, classical methodologies such as waterfall are no longer appropriate as the path toward the final goal is not always clear and can thus not be planned at the beginning. The implementation of BI requires not only a more dynamic approach, such as those enabled by Agile methodologies [1], but also the inclusion of much more diverse specialists, which may include programmers, database experts, data scientists, data engineers, statisticians, mathematicians, among others.

For these reasons, BI cannot be seen as a purely technological field despite the simplistic view adopted in Figure 9.1. It should rather be seen as the intersection of three major elements that can be organized in a pyramidal fashion (Figure 9.2): Technology, Processes, and People. Technology is the foundation as it provides the practical means for acquiring data, storing it, and using it. Technology enables the Processes that incorporate methodologies, standards, business processes, and quality metrics into one governance framework aimed at improving and ensuring BI returns. Processes, in turn, generate the information that is consumed by the People, whose main function is to take decisions according to the strategy, mission, and vision of the organization.

Given this brief introduction to some of the main aspects of BI, we can now move on to a more formal definition of the term, albeit the challenges that stem from its multidimensional nature. The term *Business Intelligence* was coined in 1989 [3] and only grew in popularity more recently, as data storage and processing hardware and tools became available. However, its core ideas already existed previously as separate components of executive information systems.

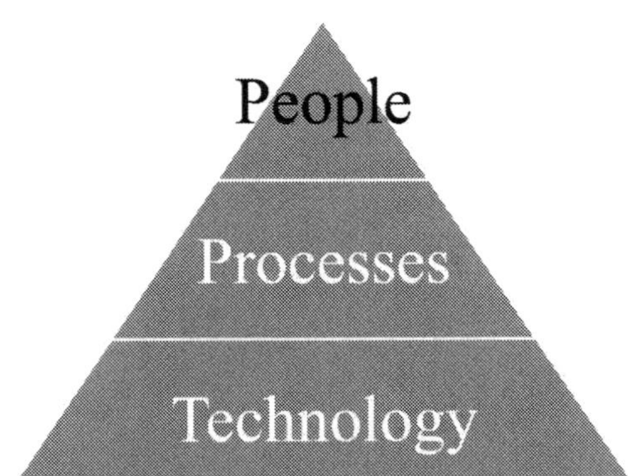

FIGURE 9.2 The three main elements of BI.

In this chapter, we look at BI as a Decision-Support System (DSS) whose main function is to consume the organization's data and transform it into valuable input for the organization's decision processes. Most of the definitions found in the literature agree on this rather broad view on BI, varying slightly according to the moment in which the definition was put forward or the perspective of the author. Specifically, three main perspectives can be identified in the literature [4]: the management perspective, the technological perspective, and the product perspective.

Howard Dresner, who is seen as the father of BI, acknowledges this multidisciplinary nature of BI in his own definition of the term, describing it as "a broad category of software and solutions for gathering, consolidating, analyzing and providing access to data in a way that lets enterprise users make better business decisions".

Similarly, Azvine et al. [5] claims that BI is about "capturing, accessing, understanding, analyzing and converting one of the fundamental and most precious assets of the company, represented by the raw data, into active information in order to improve business". Gibson et al. [6] focus its definition on the technological aspects and on the significant business value that stems from the improvements at the level of managerial decision-making effectiveness. The authors also stress the importance of these strategic information systems in todays' uncertain and highly competitive business environment.

Summarizing, BI contributes positively to several key aspects of organizational success:

- It facilitates the generation of knowledge about the organization and its domain, allowing the organization to devise solutions that better tackle current challenges and prevent future ones or at least put the organization in a better position to face them.
- Its continued use at several levels allows the organization to continuously improve as a whole, as each decision maker understands increasingly better the organization and the challenges it faces. In a sense, it can be stated that it allows the organization to collectively learn how to perform better.
- By providing more and valuable information about the organization and its challenges, BI also allows for richer decision environments, supporting creative problem-solving, and allowing the organization to device new solutions, products, or services that allow it to adapt quicker to changes in its environment.

In its present form, the BI life cycle can be seen as five main sequential phases, as proposed by [1]. The first phase is the Discovery one. In this phase, the objectives and path to follow are not yet clear for all the participants. This is, therefore, an exploratory and sometimes iterative phase in which the stakeholders identify the information requirements and the business needs, which in turn allow to create a project plan by defining the needed data sources, dimensions, facts, and technical measures. This is also a creative phase, in which the participants should be able to come up with meaningful questions given the business domain, whose answers will guide the development of the BI process. For this reason, it is also important that people with different backgrounds and with different views on the problem take part. Idea generation will also be fostered if code-free environments are used, supported by visual analysis tools, making it easier for nonspecialists to participate and contribute.

Once the business questions and the information have been made available and analyzed, the second phase takes place: Design. In this phase, the stakeholders define the architecture of the system, including not only the hardware but also components. Namely, it is in this phase that the stakeholders define the data architecture, frameworks for data presentation, the methodologies to guide the work, or the hardware and software to be used. All this should be carried out while keeping in mind the current design, its maintenance over time, and the possibility of future upgrades.

After the design phase, the actual development or implementation of the BI tool follows. This phase may include many different activities, depending on the two previous ones. Common activities include the implementation of ETL pipelines using data processing tools such as Spark or MapReduce, scripting and scheduling jobs in tools such as Cron Jobs or AirFlow, the development of dashboard or other mechanisms for data query, and visualization using tools such as Elastic Search and Kibana, among others. In the development phase, it is important that the stakeholders follow agile approaches that allow them to continuously test all the components and perform the necessary changes in due time. At the end of this phase, the system should ingest data as input and automatically analyze it and generate reports and other elements, significantly improving the efficiency of the business.

The fourth phase is the deployment one. In this phase, each component of the BI system is brought into the production environment, and it is carefully tested, both individually as well as in its integration with the remaining components. This may also include the updating of existing components, a case in which the new version must also be compared against the older one, namely to check if any existing issues have been addressed, or to check if all functionalities still work as expected. The main goal of this phase is to continuously update the BI system while, at the same time, ensuring its stability.

The fifth and final phase is the value delivery. The main goal of this phase is to ensure that the BI system continues to deliver value to the organization and, if possible, to increase this value. To this end, communication with end users is paramount as their feedback provides valuable information regarding the business needs and expectations, and how information is used. Given the continuous change in todays' businesses and the dynamic uses of information, maintaining BI value delivery requires constant attention from the stakeholders.

9.2.1 Evolution and Applications

As already addressed, the concept of BI was proposed in 1989, although some of its key ideas were already being used in many data-driven organizations. The same can be stated about the field of Analytics in general.

Data first started to be collected and used by businesses in the 1950s. At the time, data stores were relatively small, inefficient, and tended to be well structured. Data sources were few, did not produce a large quantity of data, and described mostly data that was internal to the organization. Data had to necessarily be stored (in a Data Warehouse or similar tool) to be processed, and data processing was always a batch task. Data analysts were generally not a part of the business processes as there was a clear separation between the two areas. Consequently, analytics results did not always reach the business and were not seen as fundamental for the organization. Moreover, most Data Analysis tasks were merely descriptive, and BI consisted mostly of reporting. This early version of analytics is often referred to as BI and Analytics 1.0 [7, 8].

However, the widespread use and notoriety of BI and Analytics did not reach its current levels until recently. Indeed, its recent unprecedented growth cannot be analyzed separately from the advent of Big Data. Big Data emerged in the last decade as the result of several major technological and societal changes.

First, data storage has become relatively inexpensive over the years, which allows for all sorts of data, independently of how apparently irrelevant, to be stored. Indeed, the first hard disk drive was invented by IBM in the 1950s, called the Random Access Method of Accounting and Control (RAMAC) for data storage. The RAMAC stored 5 MB of data and costed US $250,000 at the time (or $50,000 per megabyte). Nowadays, 1 TB portable hard disk drives can be bought for about US $100.

While the storage costs plummeted, IoT (Internet of Things) emerged with the promise of connecting all of our small daily devices into one global physical network. These devices constantly produce or consume data about our everyday activities, resulting in an unprecedented wealth of data characterizing us and our lives. The use of these data represents a significant opportunity for companies to get closer to the customers and their wants. The emergence of IoT was possible due to two main technological developments: the emergence of very fast networks such as 4G (Fourth Generation broadband cellular networks) and improvements in miniaturization and batteries, which allow for the development of small (eventually mobile) devices with extended autonomy.

These evolutions in the hardware were also accompanied by significant technological leaps in software. Indeed, in the last years, a wealth of tools and frameworks for processing large distributed datasets have emerged that allow for all these data to be processed either in batch or in streaming, allowing the development of real-time applications that leverage on data. Moreover, open-source technologies such as Hadoop, Mesos, or Yarn allow for large clusters of commodity hardware to be easily deployed, effectively democratizing the access to large-scale data storage and processing. On top of these clusters, many other technologies can now be used, such as Spark, Hive, MapReduce, among others, to easily store and process data.

One last interesting evolution took place in human resources. Indeed, this major technological shift led to the need for professionals skilled in statistics, data analytics, machine learning, or data engineering, who could make sense of all the data collected and derive value for the organizations. This led to the emergence of new specialists, such as Data Scientists and Data Engineers, who specialize in data exploration and transformation, and indeed fuel the whole BI process.

All these technological and societal evolutions made Big Data technologies and BI available even for small and medium enterprises that in the past might not have had the necessary resources. And they also originated what became known as BI and Analytics 2.0.

In this new era, data is mostly externally sourced and collected from many different sources. Moreover, data is collected at an unprecedented velocity. Given the competitive environment of todays' organizations, results are also required at a very different speed from those in the 1950s. This means that today, many of the data processing tasks are carried out in streaming rather than in batch (Gaber). That is, data is processed and used in real-time, without being stored first. Indeed, raw data may not even be stored at all.

Similarly, in the past machine learning models used to be trained in batch. That is, data was stored, a model was trained, and it was used by the organization from that point on. Nowadays, conditions change much faster, which means that models quickly become outdated. Models are now trained online: data arrives in streaming and models are continuously updated to prepare them for future data [9]. This means that the overall velocity at which data is processed and results are delivered is significantly faster than previously. Many visual analytics tools also emerged, which allow for information and insights to be visualized in real time.

The success and rapid evolution of BI and Analytics 2.0 is also in part explained by the role that many of today's largest companies played, including Google, Yahoo, Facebook, and more recently Netflix or Airbnb, just to name a few. These companies were among the first to leverage on really large data collection processes, and to create value from them. In the process, they faced the need to solve new challenges in data storage and processing. Many of today's technologies, most of which are Open Source, were in fact born or significantly supported by these companies. Some of the most notorious examples include Hadoop and HDFS (the Hadoop Distributed File System) that are largely supported by Yahoo and partly inspired in the Google File System, or Airflow that was created by Airbnb to manage data pipelines and later open sourced.

This is, in general, the current state of BI and Analytics. However, some authors are now drawing attention to a new evolution toward the so-called BI and Analytics 3.0. According to Chen et al. [8], the main distinctive characteristic of this new version is the highly mobile and sensor-based content. Mobility exists now not only in data collection but also in data visualization and use. As technology improves, model training and BI in general can now take place in mobile handheld devices, and be carried with us everywhere we go. This also represents a major improvement in what concerns data protection and privacy.

The focus of BI processes is also shifting as many of its applications are now person-centered and with a strong context and location awareness. Now more than ever, data about people and their habits and routines is at least just as valuable, if not more, as data about the organizations themselves.

In this new era, organizations are not so worried anymore about collecting all the data. While large clusters still exist and the Hadoop era is far from over, these are mostly used for their processing capabilities and their relative inexpensive cost from using commodity hardware. It can be argued that the Warehouse era is slowly ending to give place to a new Streaming era in which raw data are used and disposed of and only processed data and models are persisted.

As this evolution took place, analytics processes also became closer to the business processes [3]. While in the beginning there was a clear separation, many organizations nowadays incorporate the results of BI and Analytics into their business processes, automating decision-making and optimizations. When successfully implemented, this allows organizations to extract value from BI significantly faster, which may translate into a strategic advantage.

It is thus clear that the evolution of BI over the past decades, and especially in the recent past, is clearly linked to that of Big Data and related technologies. It was the current variety and velocity of the data that revealed the true potential of the field, fundamentally changing how organizations and the society use information.

Presently, BI systems are used in many different fields of application, as detailed by Chee et al. [4]. In the transportation industry, BI systems are used in areas as diverse as the Airline industry or traffic management in smart cities. In the Airline industry, BI systems allow the stakeholders to better understand traffic flows, extract knowledge from incident reports, or develop more efficient aircrafts, just to name a few. BI systems also allow to better understand or even predict traffic flows in smart cities, supporting decisions regarding road building and traffic planning.

In the Banking industry, BI systems are used by stakeholders to increase the certainty in their decision-making processes, namely in what concerns customer analysis, financial analysis, marketing, risk analysis, and fraud detection.

The Healthcare industry was also revolutionized by BI systems, especially due to the increased availability of information regarding patients, diseases, treatments, among others. Moreover, this information can now more easily be shared across systems, allowing decisions to be taken with increased quality.

Another classical application field is the Retail industry, in which BI systems are used mostly to analyze past data about clients and their transactions in order to forecast future short-term and/or long-term demand. There are also a lot of applications in Marketing, especially to promote customer loyalty.

The Manufacturing industry is another major user of BI systems, namely to monitor stocks, track inventory usage across location and time, or to track organization performance in light of the established goals.

All in all, the use of BI systems is nowadays pervasive to most of the existing industry and, in a time of very dynamic competition, it is often the difference between the success and unsuccess of the organization.

9.2.2 Challenges

Despite the generalized success of BI implementations in the industry, success is not always guaranteed when an organization begins the analysis and design of BI processes. Many of the challenges indeed arise in this first phase of conceptualization. First, the integration of end users, stakeholders, and experts with different backgrounds, each with their own conceptualizations and view of the domain, may make communication difficult. Among other activities, this may negatively impact the elicitation of requirements, which may become fuzzy and unclear. Ultimately, these communication issues may jeopardize the relationship between the different parts, causing a lack of trust between IT and business stakeholders.

Indeed, the issues that often emerge in the Discovery and Design phases may not have consequences right away, but extend and have repercussions into the following phases. Namely, sometimes the costs, time of realization, or the implementation goals are poorly estimated, which may result in the perception that the BI process does not provide enough value, or that it is too slow to do so in face of the organization's rhythm. This perception may in turn lead to the lack of a company-wide adoption, as some of its departments may feel that the effort or investment does not compensate the value provided. To prevent this from happening, the organization should have a strong company-wide BI strategy, communicated clearly to all the stakeholders, so that all are motivated to work toward its implementation and feel that they are part of the organization's data culture. It is also important that the organization defines and measures the right indicators describing the BI system in both the efficiency and efficacy dimensions.

At other times, it may happen that the implementation of a BI system is indeed too expensive and hard to justify in light of the organization's structure of dimension. This happens more frequently in small- and medium-sized enterprises (SMEs). In such cases, and when possible, the best option might be to use out of the box software rather than designing and implementing a custom solution from scratch. The cost of a BI system may also come from the necessity of providing training to the stakeholders, namely for data administration applications or for the dashboards and interfaces of the BI system. Once again, this is a more significant challenge for SMEs.

The problems pointed out in the Discovery and Design phases may also result in a lack of understanding about how data are created, what their sources are, and how it may be best used to achieve the organization's mission. This is even more challenging in domains in which there are multiple and different data sources such as different Enterprise Resource Planning, databases, or Customer Relationship Management, which may be internal or external to the organization. As a result, the value of the information provided by the BI system might be less than the expected and/or possible. In such scenarios, there are nowadays many data connectors as well as ingestion and integration tools that greatly facilitate the integration of data from different sources, such as Apache Goblin.

Another key challenge is that of data quality, which is probably the most significant predictor of the quality of the decision support provided, as further assessed in Section 4. That is, without quality data the usefulness and value of the BI system decreases significantly. Data may also be unorganized, unstructured, or it may need cleaning or curation. For these reasons, it is important that the organization implements data governance strategies, namely for measuring data quality and provenance. These may be fundamental in quantifying the quality and value of the whole BI system. The importance of Governance in the BI value chain is further analyzed in Section 3 of this chapter.

There are, however, other challenges throughout the whole life cycle. Often, the stakeholders involved may lack the insights to appropriately develop a BI process, given its specific characteristics. It may happen, for instance, that traditional and familiar development methodologies, that work in traditional software projects, are applied to implement a BI process. In these cases, the development process may lack the necessary dynamics and agility, and the ability to adapt to change. Or worse, stakeholders may not even be aware of these changes (e.g., in requirements, in information needs, in the understanding of the data) and thus fail to react accordingly and timely. As already addressed previously, the implementation of BI initiatives requires agile methodologies, the inclusion of experts from different areas, an iterative process, and the energy to continuously monitor and adapt [2].

Lately, in an epoch in which many of the decision makers are continuously moving while relying on mobile devices to work, there is also a need for mobile BI.

That is, a BI system (or at least its Human interface) that can be used on mobile devices. Here, the main challenge is the design and implementation of dashboards and other interactive elements in a way that can be easily accessed and used in the relatively small screens of mobile devices. There are, evidently, other technical challenges such as data or bandwidth restrictions.

In summary, the challenges in the implementation and delivery of BI are to enable efficient communication between the different stakeholders and the business, to acquire and transform raw data into valuable information, and to allow an easy, intuitive and technology-agnostic use of this information by the organization. If the methodology followed in the implementation of the BI system is continuously steered by these challenges, success is more likely. Sections 3 and 4 provide an in-depth analysis of these aspects and of their importance in the BI value chain.

9.3 BUSINESS INTELLIGENCE MATURITY ASSESSMENT

The implementation of BI systems requires serious investments as well as mechanisms to measure its added values. For many economic activities, the data asset represents the most valuable competitive indicator that forces companies to carefully preserve and develop systems that allow its efficient use for analyses and decision support [8]. But the knowledge becomes wisdom and meaningful information for competitive decision-making only if the existing data can be timely, efficiently, and effectively used. Many companies decide to invest significantly on BI but hardly have mechanisms to measure and calculate the ROI of that decision [10].

To measure the added business value that arises with the investments on new information systems or business models realignment, comparable references are required. Since there are no clear and globally accepted processes to measure the impact on BI adhesion [11], the use of Maturity Models (MMs) [12] can be an alternative for that purpose. Paraphrasing Rajterič [12], "... the maturity model consists of a model and questionnaire, which is used to assess the level of maturity of the development environment ...", and each level of maturity have different phases, with its own life cycle, that "...need to be completed by the organization in order to achieve a certain level of maturity ...". The MMs of any organization represent its level of preparation for the context-aware business. The stage of maturity is a practical perspective of the continuous growth, in a predictable or not, way. Since nowadays contexts are very unpredictable and instable [13], companies are forced to continuously revise and realign their strategies to keep or to improve its competitive position. Furthermore, the used MM that sustains the stability and "speed cross" of their business may require realignment too, including the life cycle of the involved phases. New versions of existing models arise, or even new models can appear.

Even generally announced as mainly applied to Information System (IS) development [14], the first steps of MM was done in the 1970s on the management of companies resources, technology, and workforce (human) dynamics [15]. The investments always "touch" persons or technology domains and continue like that till today.

IGI Global presents in Ref. [16] interesting MM perspectives: "... staged model describing different evolutionary levels towards improvement and better capabilities …"; "… set of structured levels describing how well the information and records management policies, practices, and processes of an organization have been implemented and whether they produce the required outcomes …"; "… a set of structured levels that describe how well an organization can reliably and sustainably produce required outcomes …"; "… A tool used to assess the current state of an organization in a specific context of analysis. This type of models is also used to communicate best practices and guide organizational improvements …", all essentially focused of IS development of acquisition strategies.

Assessing an MM in an organization, is equivalent to analyze the organizations' situation on intended domain and determine its compliance or not with the MM phases.

9.3.1 Maturity Assessment

Several definitions of Maturity put the term not only on the spectrum of human inherent characteristics but also on models, processes, relations and decisions. The biological perspective as "… the stage in a life cycle that is reached when a developing organism has taken on the appearance of the adult form and is capable of reproduction …" [17] or the more psychological perspective on independence, responsibility, and motivation dimensions [18] can easily be applied to organizations that are both complex business and complex social systems [19] and "living organisms in ecosystems" behaving according to models, processes, relations, and decisions. Most of the times seen as key factors for being competitive, the measurement or assessment of that level of maturity is not easy and pragmatic; however, indeed, it is a valuable resource for users and business [20].

Any assessment process should have a reference that sustains the expected grading level. The MMs are proposals for levels of maturity, which work as such references for the analysis of "… strengths and weaknesses of certain domains …" [21]. Literature sustains that this assessment strategy comes from the key role played on software development by the Capability Maturity Model (CMM) proposed by the Carnegie Mellon University (CMU) [14], and Lahrmann et al. in Ref. [22] expose the assessment approach as one of the most relevant characteristics that MM must have (Table 9.1).

Several literature contributions explore existing MM for multiple domains: Business Process [24], Business Management [25], Software development [14], IT Management [26], Project Management [27], Quality Management [28], Enterprise Architecture [26], Management Control Systems [29], Innovation [30], and many others. Since the assessment of BI initiatives is the goal of this chapter, hereafter the on-going Business Intelligence Maturity Models (BIMMs) are emphasized.

Since BI is quite related with the organizations' data and its analysis, the knowledge retrieved from their data warehouse or raw data repositories, its management, and the processes that discovery it (from descriptive to prescriptive analytics) requires complex processes and sufficiently mature models. BI, Knowledge, and Data Analysis are very closer and related concepts that are present in several levels of existing BIMMs.

Due to its relevance, several literature reviews on BIMMs [10], [12], [18], [31–33] were made. Table 9.2 presents some existing BIMM or related MMs, their main levels of maturity, and assessment key areas.

TABLE 9.1
Properties of MMs

Property	Description
Maturity concept	Three different maturity concepts: (i) People capability defines "the level of knowledge, skills, and process abilities available for performing an organization's business activities". (ii) Process maturity defines "the extent to which a specific process is explicitly defined, managed, measured, controlled, and effective". And (iii) technology maturity defines the respective level of development of a design object.
Dimension	Dimensions or domains are specific capability areas, process areas, or design objects structuring the field of interest. They should be exhaustive and distinct.
Level	Archetypal states of maturity of a certain dimension. Each level has a distinguishing descriptor providing the level's intent and a detailed description.
Maturity principle	MMs can be continuous or staged. Continuous MMs allow a scoring (aggregating operations) of activities at different levels. Staged models require the compliance with all elements of one level. They specify a number of goals and key practices to reach a predefined level. Staged MMs reduce the levels to the defined stages, whereas continuous MMs open up the possibility of specifying situational levels.
Assessment	The assessment approach can be qualitative using descriptions or quantitative using, for example, Likert-like scales.

Source: [22, 23].

TABLE 9.2
Business Intelligence Maturity Models

Model	Reference	Levels of Maturity	Key Areas
Knowledge Management Capability Assessment	[34]	Lessons Learned, Knowledge Documents, Expertise, and Data	Human capital, technological factors, knowledge life cycle, tacit/implicit/explicit nature of knowledge
Enterprise Business Intelligence Maturity (EBIM)	[33] [35]	Initial, Consolidate, Integrate, Optimize, Innovate	Process, organization, technology
Service-Oriented Business Intelligence Maturity Model (SOBIMM)	[36]	Initial, immature, controlled, managed, mature	Technology, organization, business expertise

(Continued)

TABLE 9.2 *(Continued)*
Business Intelligence Maturity Models

Model	Reference	Levels of Maturity	Key Areas
IBM Data Governance Council Maturity Model	[37]	Initial, Managing, Defined, Quantitatively Managed, Optimizing	Organizational Structures and Awareness, Stewardship, Policy, Value Creation, Data Risk Management and Compliance, Information Security and Privacy, Data Architecture, Data Quality Management, Classification and Metadata, Information Life cycle Management, Audit Information, Logging and Reporting
Gartner Data Analytical Maturity Model	[32]	Standard Reporting, Advanced Reporting, Analytics Integration, Predictive Analytics	Description, Diagnostic, Knowledge
The Ladder of Business Intelligence (LOBI)	[38]	Facts, data, information, knowledge, understanding, intuition	Technology, process, people
AMR research's business intelligence/ Performance Management Maturity Model (Figure 9.3)	[39]	Reacting, Anticipating, Collaborating, Orchestrating	Performance management, Balanced scorecard
Gartner's Maturity Model for Business Intelligence and Performance Management (Figure 9.4)	[12]	Unaware, tactical, focused, strategic, pervasive	People, processes, metrics, technology
Hewlett Package Business Intelligence Maturity Model	[40]	Business technical aspect	Business enablement, information, technology, strategy, program management
Logianalytics Analytics Maturity Model (Figure 9.5)	[41]	Standalone Analytics, Bolt-On Analytics, Inline Analytics, Infused Analytics, Genius Analytics	Data, UI, Security, Self-Services, Write and Back Workflow
TDWI's Business Intelligence Maturity Model (Figure 9.6)	[42]	Infant, Child, Teenager, Adult, Sage	Scope, Sponsorship, Funding, Value, Architecture, Data, Development, Delivery

Source: Adapted from [12, 32, 37, 43, 44].

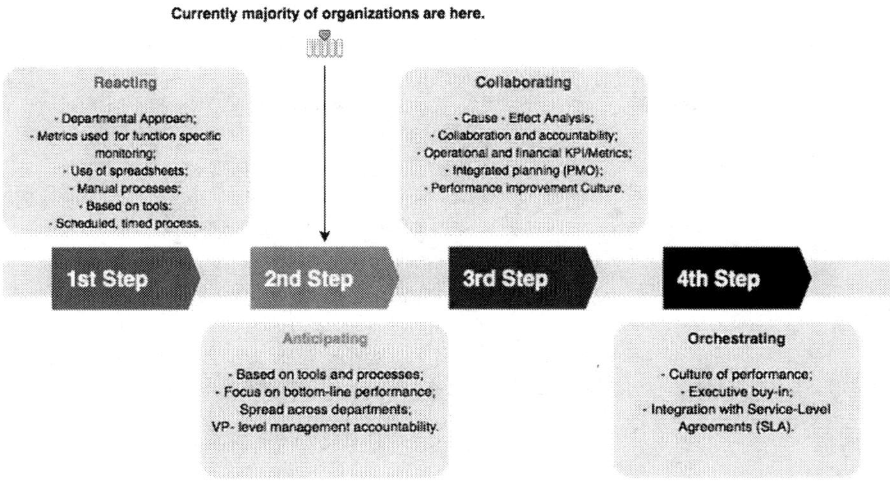

FIGURE 9.3 AMR BI/PM MM, Version 2. (Adapted from [36].)

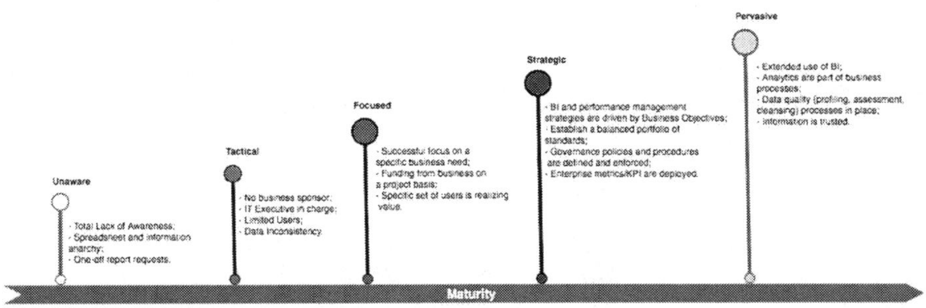

FIGURE 9.4 Gartner's BI and PM maturity model. (Adapted from [46].)

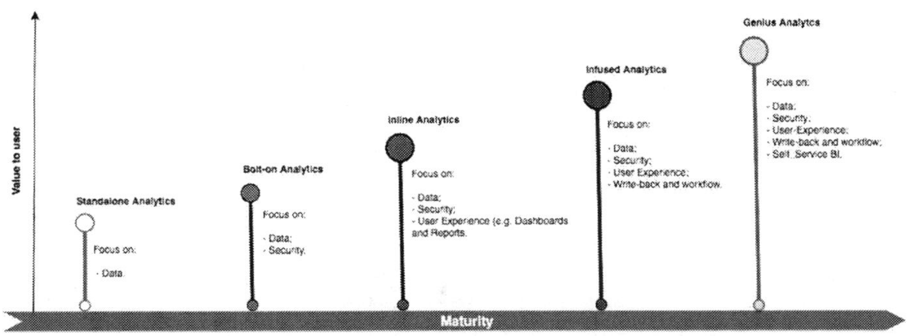

FIGURE 9.5 LogiAnalytics analytics maturity model. (Adapted from [41].)

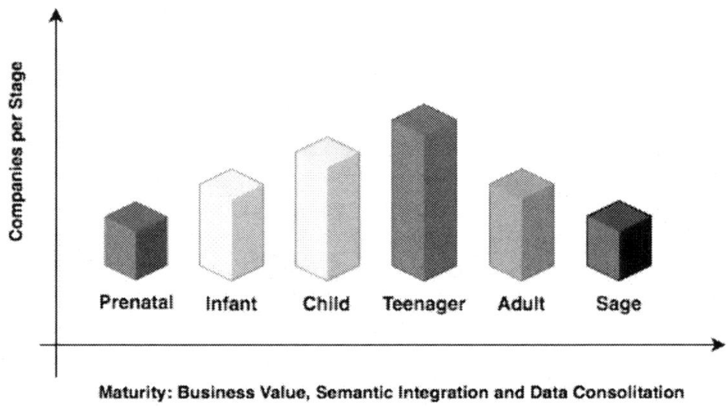

FIGURE 9.6 TDWI maturity model. (Adapted from [45].)

9.3.2 Maturity Assessment and Business Intelligence

There are several proposals of MMs to measure the capacity of entities in multiple domains (Table 9.2 presents some MM for BI). Nevertheless, the process to develop an MM is not stablished [47], being the existing proposals resulting from the experiences and strategies realignment of involved companies' business models and IT organizations. The authors of [35, 40] and [43] describe some interesting approaches and frameworks for such development processes, commonly structured in six subsequent steps (Figure 9.7).

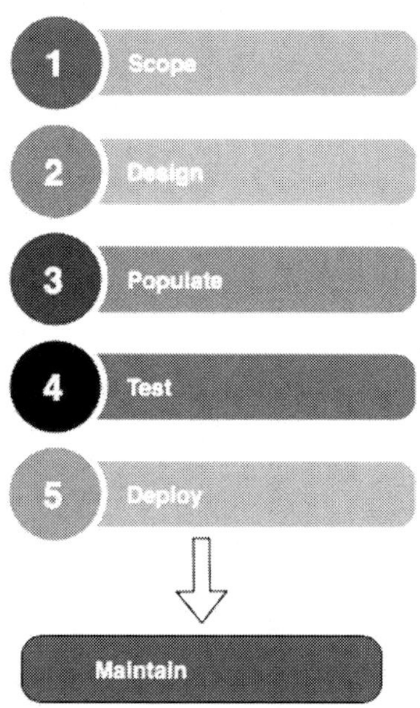

FIGURE 9.7 Model development phases. (Adapted from [43].)

The former (1) step corresponds to the initial decision to invest on BI, where ad hoc solutions or mere isolated applications support the rookie companies that just start to think about BI, or experienced restructured companies that require realignment of their strategies. This step is essential to identify the model scope and involved participants. After having the context, the second one (2) needs to measure and monitoring the business domain. Persons, processes, technologies need to be integrated in an architecture whose design will sustain the model and its dynamics for all production units, mainly ad hoc solutions, protocols, and users. The *why*, *who*, *how* are all questions that the model must be aligned (why – the relevance of the model, who and how will be involved). The third (3) step complements the model (after step 2) considering *what* needs to be measure by the model and *how* to measure it. It is essential to identify the existing domains, their participants, and their relations. The enrichment with such contents is essential to identify their interoperability and dependencies, as well as preparing for integration performance measurement. Being the model with elements to be measured and processes to support it, the next step (4) will permit the validation of the model, including obtained measurement results. Thus, validity and reliability are here explored, and the accuracy tested. Once tested, the model is prepared to be deployed and be used (5). Since the knowledge domain is dynamic the model requires to be analyzed and change accordingly (6).

The models and assessment need to be monitored and improved, for sure! Paraphrasing Tejasvi [47] on Data Governance models "(…) creating and customizing a maturity model from a strawman is effort and time incentive. But, it creates a common framework with defined capabilities, an established agreed-upon lexicon, identified objectives and deliverables as well as a list of identified evidence-based artifacts (…)" and "(…) a maturity model creates the opportunity for organizational data programs to align and demonstrate to the sponsors, the business stakeholders, the senior executives, and the oversight authorities as well as regulators that they are adhering to an industry best practice critical to building, sustaining and leveraging their data (…)".

The use of KPIs, benchmarks, or others internal or external references consubstantiates the improvement of the models or assessment methods for the BI maturity on the companies. According to Ref. [43], the maturity assessment can be (i) descriptive, having no suggestions or orientations for processes realignment or improvement; (ii) prescriptive, where paths to improvements and performance measurement are pointed to; or (iii) comparative, that essentially allows the comparison between internal and external approaches and deduce existing maturity levels. In any of those cases, the conduction of a maturity assessment process is critical for its effectiveness [48].

9.3.3 Data Governance, BI Maturity Model and Small Business

The BI on companies is quite close and dependent on existing data and its useful information. The data governance is critical and the very common dispersion of data (multiple documents, multiple types, multiple data owners) in the small companies represent serious handicaps for their BI maturity [49].

Thus, data governance MMs (Figures 9.8 and 9.9) must exist to sustain the BI maturity. Agreeing with Jan Merkus [50] "(…) increasing business intelligence maturity will make business more data-driven, and smarter and more profitable as a result (…)".

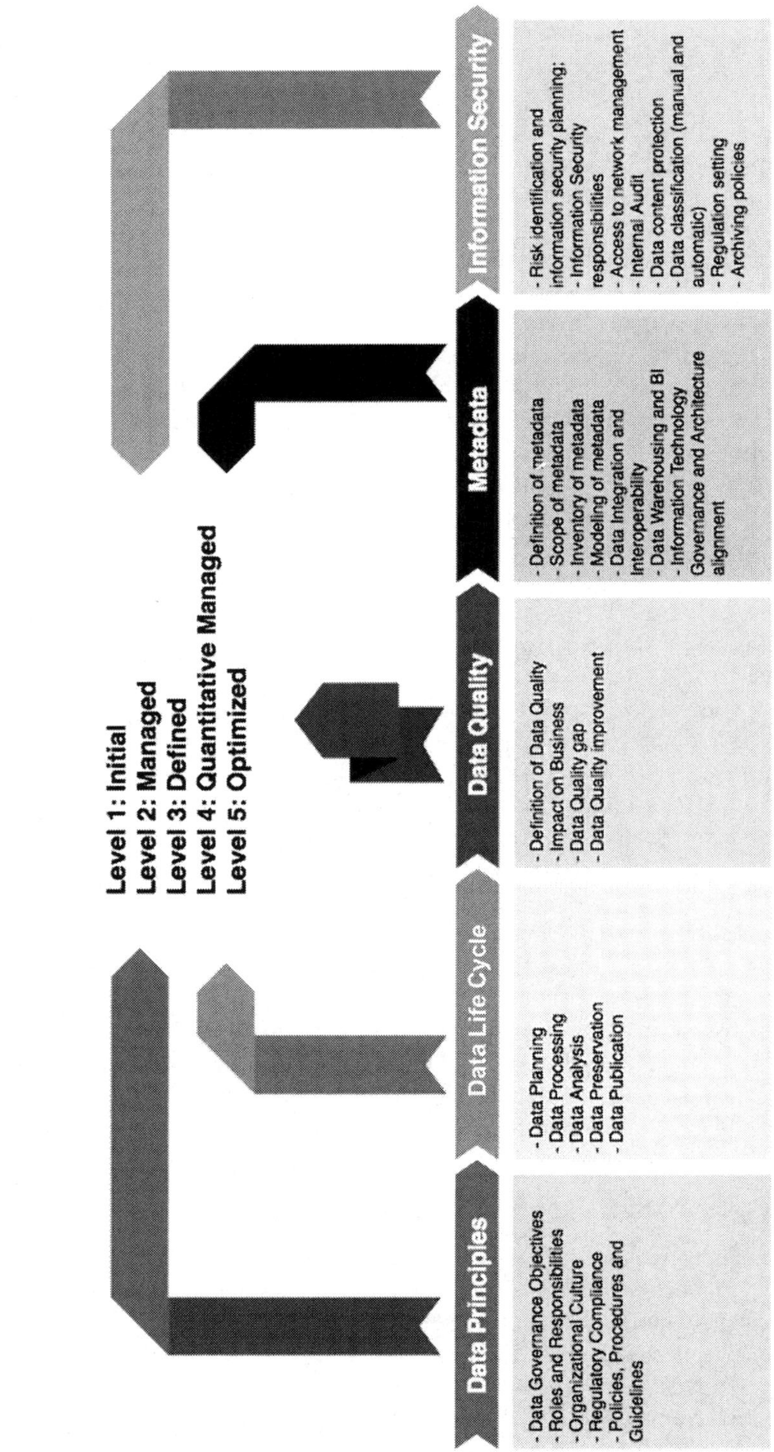

FIGURE 9.8 Data Governance Maturity Assessment Model. (Adapted from [49].)

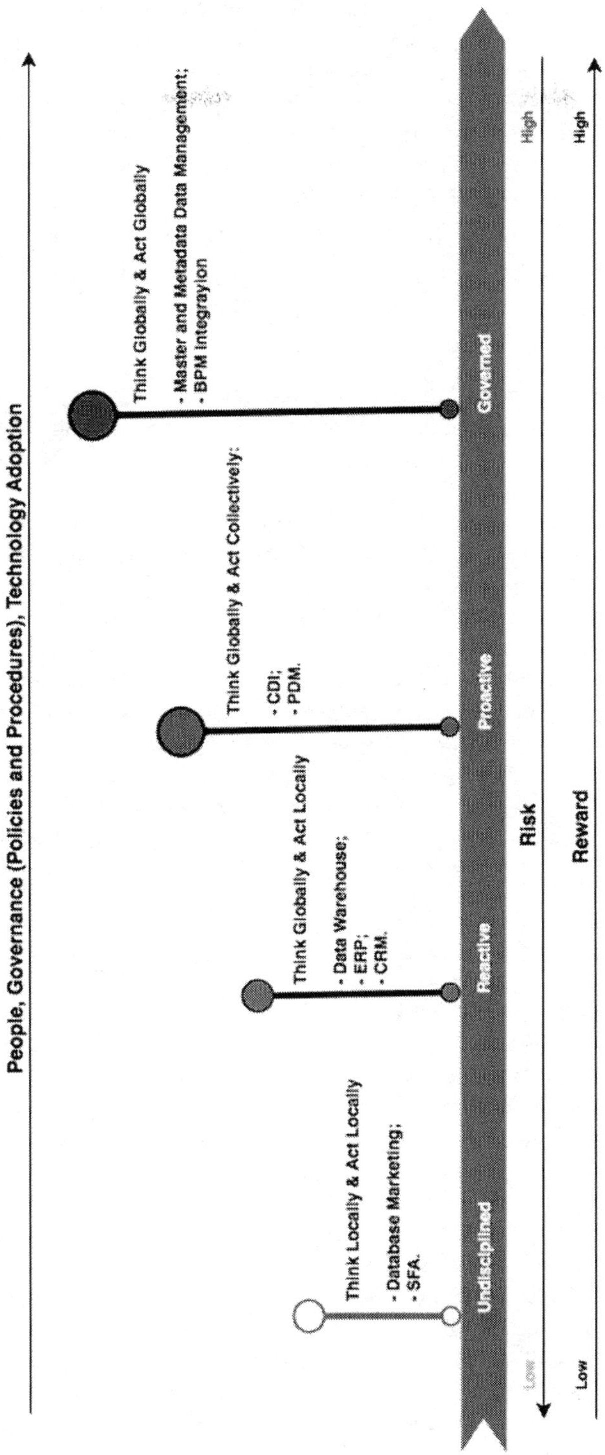

FIGURE 9.9 DataFlux data Ggvernance Maturity Model. (Adapted from [51].)

As previously referred, being people, processes and technology the main elements and concerns on any company that intends to transform its business model, the existence of dirty data represents more than an IT problem, but a business problem, instead! Policies are required to assure the perfect synchronization between all critical elements. Dataflux proposes a fourth stages process for organizations' data maturity (Figure 9.9) where "(...) each stage requires certain investments, both in internal resources and from third-party technology. However, the rewards from a data governance program escalate while risks decrease as the organization progresses through each stage (...)" [51].

9.4 DATA GOVERNANCE

As data is one of the most important and critical assets of an organization, it is really necessary to implement a set of policies, standards, and strategies for data governance for dealing with internal data management issues, but also for regulatory compliance.

Data Governance is mandatory for a successful organization to achieve master data management, improve data quality, build BI [52], being compliant with regulatory policies, ensuring that data is of high quality, it is usable, it has integrity across all the systems of the Enterprise Information System, it is protecting the privacy of the data owners, and that it is secure. Data Governance can be defined as an organizational approach to data and information management that formalizes a set of policies and procedures to guide the full life cycle of data, from collection to visualization.

The evaluation of Data Governance maturity [53] could be done using the common CMM of the CMU and the progress and impacts of a Data Governance program should be done using KPIs. The Data Maturity Model defines data management in specific process areas grouped by categories, based on the foundational principles of the Capability Maturity Model Integration (CMMI) [54]: Data Strategy, Data Quality, Platform and Architecture, Data Governance, Data Operations, and Supporting Processes.

The DAMA-DBOK2 Guide [55] identifies 11 Data Management Knowledge Areas: Data Governance, Data Architecture, Data Modelling and Design, Data Storage and Operations, Data Security, Data Integration and Interoperability, Documents and Content, Reference and Master Data, Data Warehousing and Business Intelligence, Metadata, and Data Quality.

9.4.1 DATA GOVERNANCE MATURITY ASSESSMENT

Considering the use of the CMM of the CMU for maturity-level classification, we can classify the maturity of Data Governance of the organization (i.e., degree of formality and optimization of processes, from ad hoc practices to formally defined steps, to managed result metrics, and to active optimization of the processes) in five levels: Initial, Repeatable, Defined, Managed, and Optimizing (Figure 9.10).

Considering this, we can analyze the organization data governance maturity in the following dimensions/approaches: awareness, formalization, metadata, stewardship, data quality, and master data.

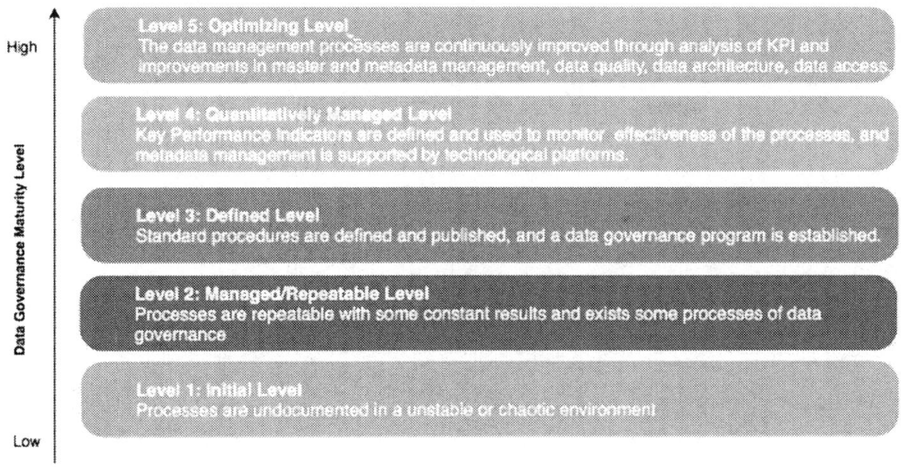

FIGURE 9.10 Data governance maturity levels. (Adapted from [53].)

So, considering that a huge group of organizations are in Initial maturity level, the starting point for these organizations included limited awareness of purpose and/per value of Data Governance, almost no well-defined data governance, and is presented in detail in Table 9.3.

Considering an organization in the initial Data Governance Maturity Level to improve the organization in data governance maturity levels, it will be necessary to create a data governance program with tasks that will impact people, processes, and technology, namely in areas such as Governance, Capabilities, Metadata Management, Master Data Management, Data Quality, and BI. This will result in a roadmap with milestones and KPIs to measure the impact and program evolution.

If the target is to move forward for the optimizing level, the characteristics are presented in Table 9.4.

TABLE 9.3
Currently Maturity Classification

Dimension	Characteristics
Awareness	Limited awareness of purpose or value of Data Governance Project.
Formalization	There are no roles related to data governance.
Metadata	Limited understanding of types and value of metadata.
Stewardship	Almost no well-defined data governance or stewardship roles or responsibilities. Data requirements are driven by the application development team.
Data Quality	Individuals perform ad hoc quality efforts as needed and manually fix data issues. Identification of data issues is based off of its usability for a specific business task.
Master Data	Inconsistent understanding of concepts and benefits of Master Data Management.

TABLE 9.4
Target Maturity Classification

Dimension	Characteristics
Awareness	Both executives and knowledge workers understand their role in the long-term evolution of the program. Knowledge workers actively promote program.
Formalization	Data governance organizational schemas are filled as defined, meet regularly, and document activities.
Metadata	A dedicated metadata management group is created to strategically advance metadata capabilities and more effectively leverage existing metadata.
Stewardship	The stewardship board includes representatives from all relevant institutional functions.
Data Quality	A data quality competency center is funded and charged with continually assessing and improving data quality outside of the system development life cycle.
Master Data	Master Data Management boards take responsibility for enforcing master data policies around their own master data across the institution.

The Optimizing Level is characterized by the following:

- Data Governance organizational schemas are filled as defined, meet regularly, and document activities.
- Compliance with official data policies is actively enforced by a governing body.
- A significant number of collaborators understand how to use data governance technologies and capabilities.
- A metadata solution provides a single point of access to federated metadata resources for structured and unstructured data.
- Multidomain master data hub handles all provisioning and management of master data.
- Compliance with Data Quality policies is tracked and reported centrally.

So, if an organization intends to increase the Data Governance Maturity Level, it needs to develop a program with activities in the following areas:

- Definition of processes and policies for organizational structure, organizational capabilities, metadata creation and maintenance; data quality and master data
- Metadata Management
- Master Data Management
- Data Quality

The implementation of a Data Governance program will increase the maturity of the company, but will also help to comply with regulations like GPDR, and sometimes the driver for Data Governance are the needs of compliance with several regulations related with data.

For example, briefly, GPDR requires that companies operating in the European Union need to have processes to monitor location and quality of data, the person accountable for that data, and the controls being applied to that data. And this is covered by data governance processes and policies, metadata management, master data management, data quality management, implementation of audit and edit log processes in the systems, implementation of rule-based access controls, and documentation of data life cycle using a data lineage. In fact data lineage is one of the foundational topics in a Data Governance program.

9.4.2 Data Governance Program Approach

A Data Governance initiative is not a project with a limited duration and should be considered a program. So, the traditional view of defining three macro phases, Design, Plan, and Execute, couldn't be seen as closed steps but as in a Scrum oriented project with consecutive execution cycles.

For the Data Governance (DG) Program should be created a DG deployment team, because experience of several projects shows that an organizational team is not formally created and this will impact negatively the progress. The most common data governance roles are Chief Data Office, Data Manager, Data Steward, Data Engineer, Business Data Owners (BDOs), organized in the Data Office under Chief Information Officer, and a Data Governance Steering Committee. The BDO reports to business areas and are in charge of business data domains.

The scope of a Data Governance program is mainly affected by three factors: business model (e.g., type of organization, hierarchy, and operating environment), systems and content being governed (e.g., data, information, documents, etc.), and degree of federation (e.g., extent or intensity by which different content is governed). This is done for example, if we can have an enterprise-wide Data Governance program or distinct business units programs. In this case, I assume we will have an enterprise wide and the most critical systems, areas, processes are the ones related with regulatory compliance: definition of data governance scope. We will need also to identify in the organization the areas related with these regulations to start work with: identify key business units/divisions.

However, for definition of priorities and roadmap of the Data Governance Program for our organization in the Maturity Level 1, one of the foundational tasks is to perform Data Lineage (starting the lineage by the most critical system and processes) because it will help to document processes, business rules, dependencies, and other data/information that will help to explain where data comes from, contributing for the creation of the metadata. And if we can implement across all areas and systems an automatic way for capturing metadata, it will be possible to include, for example,

- when data was created;
- who created the data;
- how data were acquired, added, or deleted; and
- last update of data in any data repository or system of the organization.

The impact of performing Data Lineage is transversal in the Data Governance maturity areas, contributing for

- establishment of a consistent glossary of business terms organization-wide;
- identification of duplicate date or processes;
- identification of business rules discrepancies;
- potential to identify security breaches and exposure of sensitive data;
- identification of redundant processes, systems, data, or business rules;
- compliance with regulations;
- consolidated view of data and systems architecture; and
- improved change management and new system implementation by understanding potential impacts on data flows.

The Data Lineage will obviously impact in the quality and accuracy of the data for Decision Support via Analytical Tools.

9.4.3 Tools

As enhanced previously and according to the Data Governance Maturity Levels framework, the Data Governance should be supported by tools and the adoption of following categories of tools is essential:

- Metadata Management solutions for metadata management and data lineage.
- Master Data Management.
- Data Quality.

The selection of the most adequate tool for each one of the categories identified above could be supported on benchmarks, such as Gartner Magic Quadrants [56] for the different types of platforms to identify the current ranking and then elaborate a Request For Proposals and a benchmark evaluation to choose the most adequate for the organization. One of the aspects that is important to consider in the selection of the tools to support the Data Governance program is the integration between them to operate in an orchestrated way.

A Metadata Management tool should support metadata repositories, business glossary, data classification, data profiling, data lineage, impact analysis, rules management, semantic frameworks, and metadata ingestion and translation.

The Master Data Management platform should support workflow/BPM, loading, synchronization of business services and integration, data modeling, information quality and semantics, performance, scalability, high availability and security, hierarchy management, and information stewardship support.

For Data Quality, the platform should support a high level of connectivity to huge variety of data sources, data profiling measurement and visualization, monitoring, parsing, standardization and cleaning, matching, linking and merging, multi-domain support, address validation/geocoding, data curation and enrichment, issue resolution and workflow, metadata management, DevOps environment, architecture and integration, and usability.

9.4.4 Data Governance Program Progress and Impact Analysis

For a Data Governance program, it is very important to measure the progress and impact, and this is done by the definition of KPIs (metrics), targets, periodicity of measurement, and responsibility.

The visualization of the metrics/KPI defined for the Data Governance program should be supported via Dashboards managed centrally, for example, by the Data Office department under supervision of the Data Governance Steering Committee.

It is important to have metrics to monitor the program that includes data governance maturity level, stewardship progress, definition of processes and policies, implementation and deployment of metadata management, and quality of data.

9.5 CONCLUSIONS

Currently, the industry sees the adoption, implementation, and organization-wide use of BI and Big Data tools and platforms as essential for all organizations. However, in this chapter, we argue that these approaches must be perceived as more than just a technological step, like traditional IT initiatives, because they involve many aspects other than the right tools to support Decision Support. Namely, we stress the fundamental importance of Data Maturity Assessment to evaluate how organizations are progressing and measuring Return on Investment, and Data Governance to support effective data management activities in order to accomplish business benefits but also information security, compliance, and data protection.

REFERENCES

1. D. Larson, "Agile Methodologies for Business Intelligence," in Business Intelligence and Agile Methodologies for Knowledge-Based Organizations: Cross-Disciplinary Applications. Hershey, PA, IGI Global, pp. 101–119, 2012.
2. D. Larson and V. Chang, "A review and future direction of agile, business intelligence, analytics and data science," Int. J. Inf. Manage., vol. 36, no. 5, pp. 700–710, 2016.
3. S. Negash and P. Gray, "Business Intelligence," in F. Burstein and C. Holsapple (eds). Handbook on Decision Support Systems, 2nd ed. Heidelberg, Springer, 2008.
4. T. Chee, L.-K. Chan, M.-H. Chuah, C.-S. Tan, S.-F. Wong, and W. Yeoh, "Business intelligence systems: State-of-the-art review and contemporary applications," Symp. Prog. Inform. Commun. Technol., vol. 2, no. 4, pp. 16–30, 2009.
5. B. Azvine, Z. Cui, D. D. Nauck, and B. Majeed, "Real time business intelligence for the adaptive enterprise," in The 8th IEEE International Conference on E-Commerce Technology and The 3rd IEEE International Conference on Enterprise Computing, E-Commerce, and E-Services (CEC/EEE'06), p. 29, 2006.
6. M. Gibson, D. Arnott, I. Jagielska, and A. Melbourne, "Evaluating the intangible benefits of business intelligence: Review & research agenda," in Proceedings of the 2004 IFIP International Conference on Decision Support Systems (DSS2004): Decision Support in an Uncertain and Complex World, pp. 295–305, 2004.

7. T. Davenport and J. Harris, Competing on Analytics: Updated, with a New Introduction: The New Science of Winning. Brighton, MA, Harvard Business Press, 2017.
8. H. Chen, R. H. L. Chiang, and V. C. Storey, "Business intelligence and analytics: From big data to big impact," MIS Q., vol. 36, no. 4, pp. 1165–1188, 2012.
9. D. Saad, "Online algorithms and stochastic approximations," Online Learn., vol. 5, 1998.
10. C. M. Olszak, "Assessment of business intelligence maturity in the selected organizations," Proc. 2013 Fed. Conf. Comput. Sci. Inf. Syst., pp. 951–958, 2013.
11. C. Economics, "Business Intelligence: Bright spot in IT Investment," *Computer Economics, Inc.*, 2008. [Online]. Available: https://www.computereconomics.com/article.cfm?id=1406. [Accessed: 20-May-2019].
12. I. H. Rajterič, "Overview of business intelligence maturity models," Int. J. Hum. Sci., vol. 15, no. 1, pp. 47–67, 2010.
13. G. Putnik, L. G. M. Ferreira, M. Cunha, Z. Putnik, H. Castro, and V. Shah, "User Pragmatics for Effective Ontologies Interoperability for Cloud-based Applications in Manufacturing systems," in Proceedings of the ViNOrg 2013 conference on Virtual and Networked Organizations Emergent Technologies and Tools, 2013.
14. M. C. Paulk, S. M. Garcia, M. B. Chrissis, C. V. Weber, and M. W. Bush, "Key Practices of the Capability Maturity Model," 1993.
15. C. F. Gibson and R. Nolan, "Managing the four stages of EDP growth," Harv. Bus. Rev., vol. 52, 1974.
16. M. M. Cruz-Cunha, I. M. Miranda, and P. Goncalves, Eds., Handbook of Research on ICTs and Management Systems for Improving Efficiency in Healthcare and Social Care. Hershey, PA, IGI Global, 2014.
17. E. Marti and R. Hine, A Dictionary of Biology. Oxford University Press, 2004.
18. P. Hersey and K. H. Blanchard, "Life cycle theory of leadership," Train. Dev. J., vol. 23, no. 5, pp. 26–34, 1969.
19. A. Spanyi, "Beyond process maturity to process competence." BPTrends, 2004.
20. K. Moore, "What Is a Maturity Assessment? And How to Make One in 5 Steps," *SnapApp*, 2017. [Online]. Available: https://www.snapapp.com/blog/what-how-make-maturity-assessment/. [Accessed: 23-May-2019].
21. T. Mettler and P. Rohner, "Situational maturity models as instrumental artifacts for organizational design," in DESRIST 2009, 2009.
22. G. Lahrmann, F. Marx, R. Winter, and F. Wortmann, "Business intelligence maturity models: An overview," in II Conference of the Italian Chapter of AIS (itAIS 2010), October 2010, 2010.
23. G. Lahrmann and F. Marx, "Systematization of Maturity Model Extensions," in 5th International Conference on Global Perspectives on Design Science Research, 2010.
24. J. Lee, D. Lee, and S. Kang, "An overview of the business process maturity model (BPMM)," vol. 4537, pp. 384–395, 2007.
25. T. de Bruin and M. Rosemann, "Application of a holistic model for determining BPM maturity," 2012.
26. D. Proença and J. Borbinha, "Enterprise architecture: A maturity model based on TOGAF ADM," 2017 IEEE 19th Conference on Business Informatics (CBI), pp. 257–266, 2017.
27. Z. Lianying, H. Jing, and Z. Xinxing, "The project management maturity model and application based on PRINCE2," Procedia Eng., vol. 29, pp. 3691–3697, 2012.
28. D. Y. Kim, V. Kumar, and S. A. Murphy, "European foundation for quality management business excellence model: An integrative review and research agenda," Int. J. Qual. Reliab. Manag., vol. 27, no. 6, pp. 684–701, 2010.
29. D. Frederik Marx, F. Wortmann, and J. Mayer, "A maturity model for management control systems," Bus. Inf. Syst. Eng., vol. 4, 2012.

30. R. Berg, The Innovation Maturity Model. West Palm Beach, FL, Berg Consulting Group, 2013.
31. B. Wixom and H. Watson, "The BI-based organization," IJBIR, vol. 1, pp. 13–28, 2010.
32. M.-H. Chuah and K.-L. Wong, "A review of business intelligence and its maturity models," African J. Bus. Manag., vol. 5, pp. 3424–3428, 2011.
33. W. K. Cheng and M. Rehman, "A maturity model for implementation of enterprise business intelligence systems," in Lecture Notes in Electrical Engineering, Berlin, Springer, pp. 445–454, 2018.
34. R. Freeze and U. Kulkarni, "Knowledge Management Capability Assessment: Validating a Knowledge Assets Measurement Instrument.," 2005.
35. M.-H. Chuah and K.-L. Wong, "Construct an enterprise business intelligence maturity model (EBI2M) using an integration approach: A conceptual framework," in Business Intelligence – Solution for Business Development, London, InTech, 2012.
36. E. Shaaban, Y. Helmy, A. Khedr, and M. Nasr, Business Intelligence Maturity Models: Toward New Integrated Model, The International Arab Conference on Information Technology (ACIT '11), 2012.
37. IBM, "Data governance council maturity model: Building a roadmap for effective data governance." IBM, 2007.
38. J. Cates, S. Gill, and N. Zeituny, "The ladder of business intelligence (LOBI): A framework for enterprise IT planning and architecture," IJBIS, vol. 1, pp. 220–238, 2005.
39. J. Hagerty, "AMR Research's Business Intelligence/Performance Management Maturity Model, Version 2," *Gartner Research*, 2006. [Online]. Available: https://www.gartner.com/en/documents/1345674/amr-research-s-business-intelligence-performance-managem. [Accessed: 20-Apr-2019].
40. Hewlett-Packard, "The HP Business Intelligence Maturity Model." Hewlett-Packard Development Company, 2007.
41. M. Gardner, "The Analytics Maturity Model: Five Levels of Capabilities," *Logi Analytics*, 2017. [Online]. Available: https://www.logianalytics.com/bi-trends/analytics-maturity-model/. [Accessed: 15-May-2019].
42. F. Halper and D. Stodder, "TDWI Analytics Maturity Model Guide." TWI Research, 2015.
43. T. De Bruin, R. Freeze, U. Kaulkarni, and M. Rosemann, "Understanding the main phases of developing a maturity assessment model," in Australasian Conference on Information Systems (ACIS), pp. 8–19, 2005.
44. S. Lee and L. F. Cooper, "Web 2.0 Strives to be Collaboration Software of Choice for Enterprises," *BizTechReports.Com*, 2012. [Online]. Available: http://www-01.ibm.com/software/info/itsolutions/collaboration/web20.html
45. B. W. W. Eckerson, "Beyond the Basics : Accelerating BI Maturity," 2007.
46. M. Chien, How to accelerate Analytics Adoption When Business Intelligence Maturity Is Low. Stamford, CT, Gartner Research, 2017.
47. T. Addagada, "Data Governance and the Maturity Assessment Model," *Data Top*ics, 2018. [Online]. Available: https://www.dataversity.net/data-governance-maturity-assessment-model/. [Accessed: 20-Jun-2019].
48. M. Manjrekar, "Conducting an Effective Process Maturity Assessment: du Telecom Case Study," PEX, 2012. [Online]. Available: https://www.processexcellencenetwork.com/innovation/articles/process-maturity-assessment. [Accessed: 17-Apr-2019].
49. S. Rivera, N. Loarte, C. Raymundo, and F. Mateos, "Data Governance Maturity Model for Micro Financial Organizations in Peru," 2017.
50. J. Merkus, "Data Governance Maturity Model," 2015.
51. DataFlux, "The Data Governance Maturity Model." DataFlux Corporation, 2007.
52. J. Ladley, Data Governance – How to Design, Deploy, and Sustain an Effective Data Governance Program, Amsterdam, Elsevier, 2012.

53. Data Management Maturiy, "We Need It, and How It Can propel You to Data Management Leadership", Data Management Association Phoenix, CMMI Institute, Available: http://dama-phoenix.org/wp-content/uploads/2013/07/DM-Maturity-Why-We-Need-It.pdf
54. B. C. Mary, K. Mike, S. Shrum, CMMI®: Guidelines for Process Integration and Product Improvement, Boston, MA, Addison Wesley, 2003.
55. C. Patricia, E. Susan, H. Deborah, DAMA-DMBOK2 Framework, The Data Management Association, 2012.
56. Gartner Magic Quadrant & Critical Capabilities, Gartner. [Online]. https://www.gartner.com/en/research/magic-quadrant

Index